Intelligent Electric Vehicles

Intelligent Electric Vehicles

Dr. Shaoshan Liu

SAE INTERNATIONAL®

400 Commonwealth Drive
Warrendale, PA 15096-0001 USA
E-mail: CustomerService@sae.org
Phone: 877-606-7323 (inside USA and Canada)
724-776-4970 (outside USA)
Fax: 724-776-0790

Library of Congress Catalog Number 2024943276
http://dx.doi.org/10.4271/9781468608502

ISBN-Print 978-1-4686-0849-6
ISBN-PDF 978-1-4686-0850-2
ISBN-epub 978-1-4686-0851-9

To purchase bulk quantities, please contact: SAE Customer Service

E-mail: CustomerService@sae.org
Phone: 877-606-7323 (inside USA and Canada)
724-776-4970 (outside USA)
Fax: 724-776-0790

Visit the SAE International Bookstore at books.sae.org

Publisher
Sherry Dickinson Nigam

Production and Manufacturing Associate
Michelle Silberman

Contents

Chapter 08 - Enabling Technologies—Cockpit Intelligence

Chapter 09 - Enabling Technologies - Ecosystem

Chapter 10 - Semiconductors for Intelligent Electric Vehicles

Chapter 11 - Connectivity for Intelligent Electric Vehicles

Chapter 12 - Safety Standards for Intelligent Electric Vehicles

Chapter 13 - Conclusion

Preface

Embark on a journey into the future of transportation with "**Intelligent Electric Vehicles**." This comprehensive guide is designed for a diverse audience, catering to automotive industry professionals, technology enthusiasts, policymakers, venture capitalists, and more.

What sets this book apart is its exceptional ability to demystify complex concepts, making even the most intricate technological aspects accessible to all readers. Yet, it goes beyond mere comprehension; it offers a roadmap to harness the monetization opportunities within the thriving intelligent electric vehicle (IEV) ecosystem. Enriched with real-world case studies, this book breathes life into theoretical concepts, showcasing their practical application. It does not shy away from exploring the challenges and potential roadblocks that the IEV industry faces, providing a holistic perspective.

While this book offers an extensive exploration of the IEV industry, there are certain areas it does not cover. This book does not delve into the historical aspects of the automotive industry and the evolution of traditional combustion engine vehicles. It does not provide a detailed comparison between conventional internal combustion engine (ICE) vehicles and IEVs, focusing instead on the latter's current state and future potential. Moreover, the guide does not extensively address the environmental impacts of IEV production, such as the sourcing of materials for batteries or the overall carbon footprint of vehicle manufacturing. Nor does it explore in depth the geopolitical implications of the global shift toward IEVs, such as how it affects oil-dependent economies or international relations centered around energy resources. Additionally, the book does not cover the consumer behavior aspect in detail, particularly how users are adapting to IEVs and the societal shifts associated with this adaptation.

This journey unfolds chapter by chapter, with the first five chapters covering management topics and the last seven chapters covering technical topics, so as to provide a comprehensive introduction to the whole IEV industry.

Chapter 1 highlights the wealth of monetization opportunities within the IEV industry, challenging conventional automotive business models. Chapter 2 delves into the evolution of the global IEV supply chain, revealing how it has transitioned from costly in-house development to more cost-effective strategies on a global scale. Chapter 3 expands on the supply chain topic and examines the impact of the IEV industry on our society's transition to the autonomy economy, or the use of autonomous machines, such as autonomous vehicles, delivery robots, industrial robots, drones, home service robots, etc. to provide goods and services.

Following the supply chain topic are two chapters regarding car making, where Chapter 4 presents the concept of scenario-driven car making, emphasizing customer-centric car design and offering a practical case study around autonomous mobile clinics. Chapter 5 delves into the unique project management challenges of developing IEV products.

Then, we shift gears to cover the technical topics. Chapters 6, 7, 8, and 9 are dedicated to the core technologies driving IEVs: electrification, autonomous driving, cockpit intelligence, and ecosystem. Readers will gain a profound understanding of these innovations, from cutting-edge battery technologies to the immersive world of augmented reality.

The next three chapters explore fundamental technical areas supporting the development of IEVs. Chapter 10 introduces the pivotal role of semiconductors in propelling the IEV revolution and their contribution to emerging IEV scenarios. In Chapter 11, the focus is on connectivity, exploring real-world network communication challenges and practical solutions that underpin cooperative autonomous driving. Safety takes center stage in Chapter 12, addressing the unique safety standards required for IEVs.

In summary, this book offers the following unique features:

- **Interdisciplinary approach:** The book bridges the gap between management strategy and technical knowledge, appealing to business leaders and engineers alike.
- **Case studies and real-world scenarios:** By showcasing actual applications of IEV technology, the book grounds theoretical knowledge in the real world, adding tangible value for readers.
- **Future-focused insights:** The book offers forward-looking perspectives, preparing readers for the next wave of innovations in the IEV space.
- **Coverage of emerging trends:** With chapters on cutting-edge topics like cockpit intelligence and the ecosystem of connected vehicles, the book provides insights into the future directions of the automotive industry.
- **Monetization roadmap:** The book goes beyond technical exposition to provide strategic advice on capitalizing on IEV advancements, which is particularly valuable for entrepreneurs and investors.

Foreword

As the automotive industry undergoes a profound transformation, the concept of intelligent electric vehicles (IEVs) is redefining the future of mobility. Central to this shift is the innovative approach of scenario-driven development—a method that marries cutting-edge technology with a deep, nuanced understanding of varied driving environments and consumer needs. This book, *Intelligent Electric Vehicles,* offers an in-depth exploration of this rapidly evolving field, serving as a comprehensive guide to the challenges and opportunities that lie ahead.

Scenario-driven car making embodies a meticulous focus on the diverse conditions that vehicles may encounter, such as urban congestion, rural terrains, highways, and extreme weather. By rigorously simulating these scenarios during the design and testing phases, manufacturers can craft vehicles that not only boast advanced technological features but also deliver unparalleled reliability and safety across different environments. This approach marks a significant departure from traditional car manufacturing, which often relied on standardized conditions that failed to capture the complexities of real-world driving.

In *Intelligent Electric Vehicles,* each chapter expands on this foundational concept, illustrating how scenario-driven development shapes various facets of the IEV industry. From monetization strategies and global supply chain management to the broader implications of transitioning to an autonomy-driven economy, this book examines the far-reaching impact of this innovative approach. The pivotal role of scenario-driven car making in the evolution of intelligent vehicles is particularly emphasized, underscoring its significance in the industry's future.

The book also delves into the enabling technologies that drive IEVs, including electrification, autonomous driving, cockpit intelligence, and the interconnected ecosystem that integrates these elements. Furthermore, it addresses the critical components essential for the seamless operation and safety of IEVs, such as semiconductors, connectivity solutions, and rigorous safety standards.

As we look to the future, the automotive industry's success will hinge on our ability to anticipate and adapt to the myriad challenges posed by a rapidly changing world. Scenario-driven car making offers a strategic pathway forward, ensuring that the vehicles of tomorrow are not only intelligent and electric but also finely attuned to the diverse and unpredictable realities of modern life.

Soh Wei Ming
Founder and Chairman, BeyonCa Inc
Chairman and CEO, Renault China
Founder and CEO, Mobility Asia, Volkswagen AG

Chapter 01

Monetization Opportunities for the Intelligent Electric Vehicle Industry

The best way to predict the future is to create it.

—Peter Drucker

We would like to start this book by examining the monetization opportunities for the intelligent electric vehicle (IEV) industry, such that if you want to start your career or a new venture in this emerging industry, you can pick the right area to get started.

Note that if auto manufacturing were a country, it would be the sixth largest economy in the world. As one of the world's largest industries by revenue, the automotive industry plays a pivotal role in driving economic growth, employment, and technological advancement.

The industry not only encompasses the direct manufacturing of vehicles but also engages a vast supply chain, including parts suppliers, dealerships, and aftermarket services. This expansive network makes the automotive industry a significant job creator, supporting millions of jobs worldwide, both directly within manufacturing plants and indirectly through ancillary industries such as logistics, marketing, and maintenance services.

Moreover, the automotive sector is a key driver of technological innovation and development. It heavily invests in research and development (R&D), pioneering advancements in areas such as safety features, fuel efficiency, and, more recently, autonomous driving technology and electric vehicles (EVs). These innovations not only enhance the industry's product offerings but also spill over into other sectors, fostering broader technological growth.

Additionally, the industry plays a crucial role in the development of infrastructure, as the demand for vehicles influences urban planning and the expansion of road networks. However, it is important to acknowledge the environmental challenges associated with the automotive industry, particularly in terms of emissions and resource consumption. The shift toward EVs and more sustainable practices illustrates the industry's response to these challenges and its potential to contribute positively to environmental goals. As such, the economic impact of the automotive industry is multifaceted, driving job creation, technological advancement, and infrastructural development, while also adapting to the evolving demands of environmental sustainability.

In 2022, the global automotive manufacturing industry generated a total revenue of approximately $1875.5 billion. This figure represents a modest compound annual growth rate (CAGR) of 0.8% from 2017 to 2022. The industry encompasses the production of trucks, passenger cars, and motorcycles, with the Asia-Pacific region being the dominant contributor to this market value in 2022. In 2017, global direct employment in the industry was estimated at nearly 14 million workers.

The traditional automotive industry generates most revenue and profits by selling vehicles and maintenance services. Compared to the traditional automotive industry, the IEV industry opens a new door to numerous monetization opportunities. We summarize the monetization opportunities in the IEV industry [1]. For each monetization area listed below, we compare the differences between traditional automotive business and IEV business, so as to identify the potential business opportunities for IEVs.

Note that although IEVs are products by themselves, they are more importantly platforms for emerging business models. We attempt to summarize an initial list of monetization opportunities of the IEV industry, and there are still many monetization opportunities that are yet to be defined.

- **Car Sale:** Car sale is the most profitable business model for traditional automotive original equipment manufacturers (OEMs). In 2021, global car sales grew to around 70 million units and led to a $2.86 trillion market size. Today, about 10% of the market belongs to IEVs, leaving tremendous headroom for future growth.
- **Used Car Sale:** The global used car market size was $1.50 trillion in 2021, but only a negligible percentage belongs to IEVs. Since IEVs are equipped with better telemetric data tracking capability, it is easier to track and value used IEVs. Leveraging the data advantages over traditional cars, an IEV used car marketplace is a hidden monetization opportunity that is yet to develop.
- **After-Sale Services and Maintenances:** The vehicle maintenance and services market size reached $800 billion in 2021. According to American Automobile Association (AAA), it is actually more affordable for IEV maintenance compared to nonelectric vehicles. The maintenance for each IEV costs around $949 a year compared to $1279 a year for a nonelectric vehicle. IEV OEMs have access to most of the data generated by each IEV. Because of their data advantage, maintenance can be another major revenue stream for IEV OEMs.
- **Car Accessories:** The global automotive accessories market reached $500 billion. The connectivity and computing power available in IEVs enable these IEVs to integrate more interesting accessories and gadgets, such as augmented reality (AR) glasses or virtual reality (VR) headsets. I expect a rapid growth of the IEV accessories market in the next

decade. This creates a highly profitable area for tech startups.

- **Insurance:** The global auto insurance market size is around $750 billion today. Because IEVs can capture drivers' driving behaviors as well as the vehicle's detailed operational data, there is an opportunity to revolutionize the auto insurance industry and greatly improve the efficiency, and thus gross margin, of the auto insurance industry. The key question is whether existing insurance can get access to these data or whether each IEV OEM should run its own insurance branch as well.
- **Charging:** Charging is a new business model for IEVs, and it has been expected that charging will present a huge business opportunity, projected to reach at least $15 billion by 2030. Like the gas station business, charging stations come with additional business opportunities, such as coffee, delis, car wash, and other services. At this point, it is still too early to estimate the eventual size of the charging station market size for IEVs or how these charging stations should be operated to create more monetization opportunities.
- **Car Data:** IEVs generate enormous amounts of data, as much as 2 GB/s. These pieces of data present a golden opportunity for monetization, a new area that can soon reach a market size of $400 billion. This is exactly how Internet companies, such as Google and Facebook, generate most of their revenue. IEV OEMs need to equip themselves with the necessary technologies in order to monetize the ever-growing user data. Probably in the near future, each IEV OEM will have an internal car data service provider (CDSP) group responsible for creating a new revenue stream from data.
- **In-Car Entertainment:** The global video streaming market size reached almost $400 billion and the global gaming market reached $235 billion in 2021. In the future, many of the video streaming and other forms of entertainment will take place in IEVs. Particularly, contents specially developed for in-car consumption, such as short videos, will have a unique growth opportunity (Figure 1.1).
- **Location-Based Services:** The global location-based services (LBS) market size is around $50 billion. IEVs and LBS are natural fits to each other, and LBS will become an essential monetization engine for IEVs. I believe the LBS opportunity lies in conversational LBS, where services are recommended to users through conversations.
- **Autonomous Driving Packages:** We are still at the dawn of autonomous driving monetization. Only recently, Tesla started to offer autonomous driving packages at $15,000 per license. With an increasing number of IEVs sold, autonomous driving packages sale can potentially become a $100 billion business in the future, assuming that over 10 million vehicles will activate autonomous driving services.
- **Emerging Ecosystems:** There are many emerging applications for IEVs, such as autonomous mobile clinics (AMCs), which are IEV fleets with autonomous driving capabilities that can be utilized to provide healthcare services to solve the healthcare access problem [2]. AMCs are self-driving vehicles equipped with medical diagnostic equipment, telemedicine capability, and artificial intelligence (AI) software that can provide healthcare services. Today, the global outpatient services market size is over $3 trillion. With AMCs, most of the health monitoring and outpatient services can be provided more efficiently and effectively. Concepts similar to AMCs, such

as autonomous mobile offices, autonomous mobile restaurants, autonomous mobile hotels, etc., can also be developed to transform each of these trillion-dollar industries.

- **Semiconductor:** The global automotive semiconductor market has shown significant growth and is projected to continue expanding in the coming years. In 2022, the market size was valued at around USD 59.7 billion and is anticipated to reach USD 153.9 billion by 2032, growing at a CAGR of 10.1% from 2023 to 2032 [3]. The market's growth is driven by various factors, including the increasing electronic components in vehicles, rising adoption of EVs, advancements in autonomous driving, and growing demand for enhanced vehicle safety features. Additionally, stringent regulations on vehicular safety and the surge in demand for advanced vehicle safety and comfort systems are contributing to the market expansion. The Asia-Pacific region, particularly China, is playing a significant role in the market's growth, accounting for a substantial share due to its position as a major automotive hub.
- **Connectivity:** The automotive connectivity market is experiencing significant growth and is projected to continue this upward trajectory in the coming years. As of 2023, the market is valued at approximately USD 33.42 billion and is expected to reach USD 190.29 billion by 2033, advancing at a CAGR of 19% [4]. A key driver for this market is the increasing demand for connected vehicle technologies. For instance, in China, more than 21.48 million passenger cars were sold in 2021, indicating a strong market for predictive technology in automobiles. Various segments within the automotive connectivity market are showing promising growth. The driver assistance segment is anticipated to dominate

Figure 1.1 In-car entertainment in a smart cockpit.

Gorodenkoff/Shutterstock.com.

the market due to the increasing implementation of advanced features such as adaptive cruise control, lane keep assist, and park assist, which enhance vehicle safety. The mobility management segment, which offers vital information like weather and road conditions, is also holding a significant position in the market. Additionally, the vehicle telematics segment is expected to see considerable growth, driven by the improving IT infrastructure along highways.

In summary, the IEV industry is evolving into a platform for encapsulating complex business models, including business-to-business models for providing mobility services, business-to-consumer models for vehicle and service sales, subscription-based models for in-vehicle software packages, and on-demand business models for in-vehicle entertainment. In addition, IEVs may disrupt other sectors such as healthcare. As this IEV business ecosystem matures, I project that the IEV industry will expand its market size to tens of times the size of the traditional automotive industry.

References

1. Liu, S., "Monetization Trends and Predictions for the Intelligent Electric Vehicle Industry," *Forbes*, accessed June 1, 2024, https://www.forbes.com/sites/forbestechcouncil/2022/11/10/monetization-trends-and-predictions-for-the-intelligent-electric-vehicle-industry/.

2. Liu, S., Kong, A., Huang, Y., and Liu, X., "Autonomous Mobile Clinics," *Bulletin of the World Health Organization* 100, no. 9 (2022): 527.

3. Allied Market Research, "Automotive Semiconductor Market," accessed June 1, 2024, https://www.alliedmarketresearch.com/automotive-semiconductor-market-A14477.

4. Future Market Insights, "Automotive Connectivity Market Outlook (2023 to 2033)," accessed June 1, 2024, https://www.futuremarketinsights.com/reports/automotive-connectivity-market.

Chapter 02

Managing a Global Supply Chain

The most important objective of a supply chain is to deliver the right product to the right place at the right time—and at the right cost.

—Anonymous

In this chapter, we embark on a journey through the intricate supply chains that fuel the IEV revolution. This chapter delves into the transformation of these supply chains over the past decade, highlighting the pivotal role of key players such as China and the Association of Southeast Asian Nations (ASEAN). From the evolution of the IEV industry's core components to the governmental support and technological incubation that have propelled its growth, we navigate through the intricate web of factors that define the IEV supply chain.

2.1. Evolution of Supply Chains from Traditional Automobiles to IEVs

The supply chain for IEVs differs significantly from that of traditional internal combustion engine (ICE) vehicles, primarily due to the distinct components and technologies involved as listed in Figure 2.1.

Figure 2.1 An internal combustion engine and an electric vehicle motor.

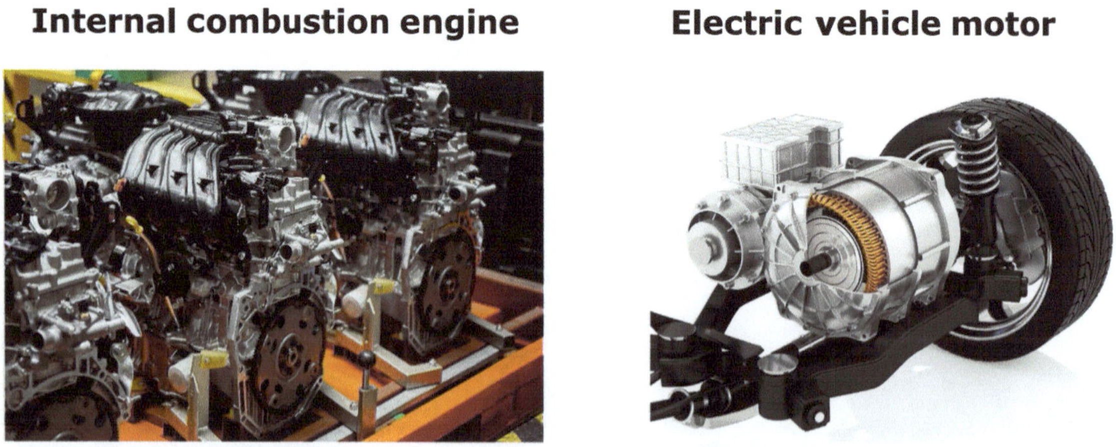

Nordroden/Chesky/Shutterstock.com.

First, IEVs are mechanically simpler than ICE vehicles. For instance, the electric motors in EVs have far fewer components compared to a traditional ICE. A comparison between the Chevrolet Bolt's electric engine and a four-cylinder ICE found that the electric motor had three moving parts compared to the ICE's 113. This simplicity impacts the number and type of components required in the supply chain [1].

Second, while mechanically simpler, IEVs are more complex in terms of electronics and software. The value of a car's powertrain and electronics is expected to rise significantly, accounting for a larger share of the car's overall value. This shift demands a supply chain that can support advanced electronic components and sophisticated software [2].

Third, the battery systems and electric drives are critical for IEVs and represent a significant portion of the vehicle's value. The shift from ICEs to IEVs demands a transformation in the supply chain to accommodate the production and distribution of these components. Battery production, in particular, is a major focus area.

Fourth, for autonomous IEVs, components like light detection and ranging (LiDAR) sensors, radar sensors, and advanced computing hardware are essential. These technologies form a significant part of the future automotive supply chain, reflecting the industry's shift toward automation and connectivity.

Fifth, IEVs are generating massive amounts of data due to the rise in connected sensors and devices. This adds layers of complexity to the supply chain, requiring robust data management and cybersecurity measures [3].

2.2. Key Components in the IEV Supply Chain

The IEV industry has progressed greatly in the past decade. Particularly, the IEV supply chain has matured and many new suppliers have emerged [4]. In the early days of the IEV industry, IEV OEMs had to do almost everything in-house.

When Tesla started its IEV business, it had to invest billions of dollars into battery systems, electrical and electronic (E/E) architectures, autonomous driving capabilities, and intelligent cockpits. Similarly, the pioneers of China's IEV industry, NIO, XPeng, and Li Auto, all spent billions of dollars developing their first vehicles due to the immaturity of IEV supply chains. It has been a common understanding that it would cost at least a billion USD to develop a new IEV five years back. With the evolution of the IEV supply chains, today developing a new IEV could cost as low as 200 million USD.

One recent trend is that, after pouring billions of dollars into their products, these IEV pioneers are all on the move to own the whole IEV supply chain in order to consolidate their established advantages in the IEV market. For instance, China's BYD is actively investing in Indonesia to secure battery supplies. Tesla has developed its proprietary computing system instead of relying on external computing solution suppliers.

Does this mean there would be no more opportunities for newcomers? I argue otherwise. I believe that the evolution of the IEV supply chains enables a more efficient and effective division of labor. Although established IEV OEMs have invested heavily in electrification and intelligence, including battery systems, computing solutions, and autonomous driving, for IEV newcomers, instead of developing everything in-house, they can leverage the advantages of the ever-maturing IEV supply chains and focus their financial resources on critical differentiation areas, such as usage scenarios, product designs, and customer engagements. The following key development trends of the IEV supply chains keep the windows of opportunity open for IEV newcomers.

Skateboard Chassis: Electrification is the fundamental technology for IEVs, and traditionally, chassis development is the core competency for automotive OEMs. The emergence of skateboard chassis kills two birds with one stone, solving both the electrification and chassis development problem simultaneously.

A skateboard chassis includes a base platform to integrate the batteries, electric motors, and other electronic components. It also embeds key automotive components such as the wheels, the suspension, the steering, the powertrain, and braking functions. The availability of skateboard chassis greatly cuts down the cost and complexity of IEV engineering as well as the manufacturing of IEVs. For instance, Chinese battery manufacturer CATL recently announced their skateboard chassis integrating CATL's battery packs, electric motors, and other components, transforming CATL from a pure electrification supplier to an electrification and chassis supplier.

With skateboard chassis, IEV OEMs can focus on product design. Instead of building a 1000-person car engineering team, a 100-person team could be sufficient to integrate a chosen skateboard chassis into the product engineering process.

Computing Solutions: Many established IEV OEMs attempt to deepen their technological moat through developing proprietary computing solutions. For instance, China's NIO has assembled a team of 300 engineers to develop self-driving chips. Similarly, XPeng has an internal team of over 200 engineers developing a self-driving chip with high computing power similar to Tesla's Full Self-Driving (FSD).

However, the semiconductor is an investment-intensive business, and as major semiconductor firms, such as Intel, NVIDIA, Qualcomm, and Texas Instruments (TI), enter the autonomous driving business and gradually roll out mature computing solutions, I am not sure that the internal semiconductor units of IEV OEMs can outcompete these established semiconductor firms. For IEV newcomers, it would be wise to directly integrate existing computing solutions instead of heavily investing in this area.

Autonomous Driving: Inspired by Tesla's success in autonomous driving, many IEV OEMs have built large internal autonomous driving teams to develop autonomous driving technologies from scratch. For instance, Tesla's autonomous driving team consists of more than 1000 engineers. Similarly, China's NIO, XPeng, and Li Auto each have autonomous driving teams exceeding 500 engineers.

Even the IEV newcomer Jidu Auto has an autonomous driving team of over 200 engineers although Baidu is developing the full autonomous driving solution for them. Another IEV newcomer in China, Xiaomi, has already built a 500-person autonomous driving team before their products have been defined.

Although it is great to have in-house autonomous driving capability, a large autonomous driving team also imposes a huge financial burden. I believe the optimal team size for autonomous driving should be around 50 engineers, who can focus on developing scenario-based autonomous driving upon existing mature autonomous driving solutions from autonomous driving technology suppliers. Reinventing the wheel with a large internal team not only incurs unnecessary financial costs but also the opportunity costs of missing the launch timeline.

In summary, although established IEV OEMs have launched many new products to market, the variety of IEVs still pales in comparison to traditional ICE vehicles. We are still at the dawn of the IEV era. Particularly, the evolution of the IEV supply chains enables IEV newcomers to develop their products cost-efficiently and time-efficiently. This is still a golden era for IEV startups as long as these startups define their products clearly and spend their financial resources efficiently. My estimate is that with a 200 million USD investment, a team of 300 experienced engineers, and 18 months of time, a new IEV can be developed.

2.3. China's Bet on IEVs

As we've explored the evolution of the IEV supply chain, it becomes evident that the industry is poised for significant growth and transformation. Now, let us shift our focus to China's strategic bet on the IEV sector and how it has positioned itself as a global leader in this burgeoning industry.

In the past two decades, the real estate sector has been an internal engine of China's economic growth. In 2021, real estate investment reached RMB 14.8 trillion, accounting for 27% of China's fixed asset investment.

However, this overextended sector has lost steam and brought China into a systematic economic problem. Realizing its overreliance on the real estate sector, in the past decade, China has made a huge bet on the IEV industry, in the hope that the IEV industry, along with its fast-growing ecosystem, will be the new engine for China's economic growth.

In this section, I explain why China has gained a competitive advantage as well as how the Chinese government has acted as a technological incubator, and project how the IEV industry will reshape the global economic landscape [5].

In 2021, the global automotive industry enjoyed a market size of $2.86 trillion from car sales alone, with only 10% of the market belonging to IEVs, leaving tremendous headroom for future growth. Historically, European, US, and Japanese car makers have dominated the ICE automotive market, both economically and technologically. Today's rapid transition from ICE cars to IEVs presents a once-in-decades opportunity for newcomers, similar to the rise of smartphone manufacturers and the fall of feature phone manufacturers in the 2010s.

Technologically, the IEV industry has been evolving through three stages of development, **electrification, intelligence, and ecosystem**: Electrification is the process of powering the vehicle with electricity, of which battery technology and supply chain are the key enablers. Upon the foundation of electrification, intelligence, such as autonomous driving, can be infused into IEVs to make the driving and riding experiences more convenient and enjoyable. Computing chips, AI talents, and data are key enablers of intelligence. Powered by intelligence and enormous computing capabilities, an application ecosystem can be developed on IEVs, similar to how mobile applications drive the sale and evolution of smartphones. With an IEV ecosystem developed, many usage scenarios as well as business opportunities beyond mobility will emerge.

Road to Electrification: Since China became a member of the World Trade Organization (WTO) in 2001, China has been a major global supplier of electronic components. In particular, battery production was considered a high-pollution industry. Many Western economies moved battery production overseas, and China became a perfect destination due to its low labor cost and relatively high tolerance to high-pollution industries back then.

This reshaping of the global electronics supply chain led to the rise of Chinese battery companies such as ATL, and later CATL, as well as BYD. In the 2010s, the ubiquitous proliferation of smartphones gave Chinese battery companies another push. Particularly, fierce competition to get into the iPhone supply chain forced these Chinese battery companies to increase their investments in technologies and product quality. Benefiting from globalization and the smartphone era, now that the IEV era is upon us, Chinese battery companies, exemplified by CATL and BYD, are already dominating the global battery market. The advantage in electrification sets the stage for China to bet on the IEV industry.

Government as a Technological Incubator: Building upon a mature electrification supply chain, the Chinese government has committed an enormous amount of financial resources and played a crucial role in incubating China's IEV industry through providing incentives for both the supply and consumption of IEVs.

To boost IEV consumption, since 2010 the central government has initiated a consumer subsidy program, which covers over 40% of the costs of IEVs. Since 2014 all IEV sales have been exempt from purchase taxes. To encourage the supply of IEVs instead of ICE cars, China has instated a "dual-credit policy" since 2018. Under the policy, automotive manufacturers are assessed in terms of fuel consumption and EV production in order to qualify for new energy credits.

As a result of these efforts, China not only owns the most sophisticated IEV supply chain but is also the biggest IEV market in the world. The natural next step for China's IEV industry is globalization, as exemplified by NIO's recent effort to enter the European market, and BYD's success in the ASEAN market. Due to their superior products, we can expect that Chinese IEV brands will soon become global brands.

Trickle-Down Supply Chain: The IEV industry is the crown jewel of the broader automation sector, which includes service robots, industrial robots, delivery robots, autonomous drones, etc. The automation sector can potentially ease China's aging population and labor shortage problem and have a much greater impact on the global economy (Figure 2.2).

Figure 2.2 Humanoid robots.

Phonlamai Photo/Shutterstock.com.

A subtle implication of the rise of China's IEV industry is that many IEV technologies and supply chains can trickle down to fuel the development of various types of autonomous machines. This is a natural move for any IEV company. Recently, Tesla announced Tesla Bot, a conceptual general-purpose robotic humanoid that can act as a home service robot or an industrial robot. Tesla Bot was made possible by Tesla's mature AI technologies as well as its manufacturing supply chain. China's XPeng made a similar move to utilize its technologies and supply chain to produce a robot pony for delivery tasks. We will further explore this topic in the next chapter.

In summary, globalization equipped China with a sophisticated manufacturing supply chain, especially in electrification. After a decade of continuous investment, China is now poised to take on the global leadership of the IEV industry. With the IEV technologies and supply chains trickling down to the broader automation sector, China is creating its economic growth engine for decades to come.

Nonetheless, before China's bet pays off, there are a few caveats. First, one core technology in the IEV supply chain is semiconductor, which China today still relies on external suppliers. Thus, any embargo on chip supply will be devastating to China's IEV industry. Second, due to the current geopolitical situation, many countries may ban Chinese IEV products over data privacy and national security concerns, similar to how the United States has banned Huawei's products.

2.4. The Next Frontier of the IEV Industry

Having delved into China's strategic endeavors in the IEV industry, we now turn our attention to the next frontier, the ASEAN, and examine why this diverse region may hold the key to the future of the global IEV industry.

ASEAN consists of ten member states, including two developed countries, Brunei and Singapore, five developing countries, Malaysia, Indonesia, Thailand, Philippines, and Vietnam, and three least developed countries (LDCs), Cambodia, Laos, and Myanmar. Despite its current economic situation, ASEAN consists of 8.58% of the world's population, and its economic growth outpaces the world's economic growth.

Although today the United States and China are leading the global IEV industry, in this section, through analyzing ASEAN's economic and political situation, I argue that ASEAN will become the next frontier of the global IEV industry [6].

We have observed tremendous growth of the global IEV market in the past few years. In 2021, EV sales represented 9% of all passenger vehicle sales. The main drivers in this emerging market are mainly the US and China automotive OEMs, with the United States's Tesla owning 21% of the battery-electric-only EV market, China's SAIC owning 13%, and China's BYD owning 7%.

On the consumption side, China and the United States are major EV markets; thus, most EVs manufactured in the United States and China can be consumed locally. On the supply side, both China and the United States have advanced manufacturing capabilities and have established global supply chains. On the infrastructure side, China and the United States have provided significant incentives to boost the development of the EV industry.

However, with the recent worsening US–China relationship, it might be increasingly challenging for Chinese IEV OEMs to take market shares in the US and European markets, and vice versa. Even worse, constraints on key components such as batteries and electronic control units (ECUs) may impose challenges on the supply side of EV products.

Given the geopolitical situation, ASEAN might be a new frontier for the IEV industry. ASEAN thus far has stayed neutral amid US–China tensions. ASEAN does not have to face the challenges of sourcing essential components throughout the world, and ASEAN does not have to worry about shipping its products to other parts of the world. These reasons make ASEAN a great region to produce EVs for the whole global market.

But is ASEAN ready to take the lead in the emerging IEV industry? First, let us examine the consumption side. After all, a strong local market serves as the basis for incubating a new industry. In the past decade, ASEAN countries have experienced rapid economic development. With the recent US–China tensions situation, many European and US businesses have migrated from China to ASEAN countries, giving a boost to ASEAN foreign direct investment (FDI), which will eventually lead to an improvement in buying power in ASEAN. In particular, exemplified by the recent VinFast success story, it has become clear that there exists great potential for the IEV market in ASEAN and that ASEAN demands its own EV brand.

Second, on the supply side, there are four key components in IEV manufacturing, software and AI talents, batteries, ECUs, and manufacturing capabilities. Interestingly, ASEAN as a whole actually has all these components ready. Singapore, powered by its superior university and research and development (R&D) systems, has been a software and AI talent powerhouse. Indonesia is extremely rich in natural reserves of nickel and cobalt, which are essential battery materials, making Indonesia a highly competitive battery supplier. In the past two decades, Malaysia has positioned itself as a major semiconductor manufacturer in the global semiconductor supply chain, especially in the automotive ECU space. Last but not least, with manufacturing plants of Toyota, Isuzu, Honda, Nissan, General Motors, Ford Motor, etc., the automotive industry in Thailand is the largest in Southeast Asia and ready for EV manufacturing.

Third, on the infrastructure side, many studies have concluded that insufficient charging infrastructure has been the main reason why ASEAN consumers are not choosing EVs. For instance, there are only 187 units of charging stations throughout Indonesia, compared to over 1.15 million charging poles in China. To really boost the development of the IEV industry in ASEAN, ASEAN governments should really follow in the footsteps of China and the United States, to invest in the charging infrastructure and to provide incentives for people switching from gasoline-powered vehicles to EVs.

Fourth, while building the IEV ecosystem in ASEAN, many innovations and business opportunities will arise along with the ecosystem. For instance, new in-car entertainment applications will arise along with the EV industry. Particularly in the past few years, many highly successful startups, such as Grab, Sea, Jobstreet, and Go-Jek, have emerged in ASEAN, gradually cultivating an entrepreneurship culture within the region. The ASEAN governments should invest in incubating startups around the IEV ecosystem, so as to solidify ASEAN's global leadership position in this industry. China has achieved it in the mobile Internet era, maybe it is ASEAN's turn in the IEV era.

Major automotive OEMs around the world should definitely invest in ASEAN to capture this once-in-decades market opportunity. As many companies are already investing in ASEAN to secure resources and market share. The latest examples are China's BYD investing in Indonesia to secure battery supplies, and China's NIO making a push to penetrate the ASEAN market.

In conclusion, due to its recent rapid economic growth, ready supply chain, and the current geopolitical situation, I believe ASEAN is poised to become the next frontier of the IEV industry. To incubate this emerging industry, the ASEAN governments should invest in charging infrastructure and encourage entrepreneurship. But even before ASEAN is ready, major players in the IEV industry should invest in ASEAN to secure resources and market shares, before the market gets overcrowded.

2.5. Summary

In this chapter, we explore the dynamic landscape of supply chains driving the IEV revolution. We delve into the evolution of IEV supply chains, highlighting their departure from traditional automobile supply chains due to differences in components, electronics, and software complexity. Key findings reveal the shift toward simpler mechanical structures but increased complexity in electronics and software, with battery systems, electric drives, and autonomous driving components emerging as critical elements.

Despite the challenges, the IEV supply chain has gradually matured, allowing for the cost-efficient development of new IEVs. Established players are consolidating their supply chains, but opportunities remain for newcomers to leverage matured supply chains and focus on differentiation areas such as product design and customer engagement.

As the IEV industry matures, China emerges as a global leader, backed by a sophisticated supply chain and government support. However, challenges such as semiconductor dependence and geopolitical tensions loom. Looking beyond China, ASEAN presents a promising frontier for the IEV industry. With its growing economy, ready supply chain, and geopolitical neutrality, ASEAN offers significant potential for IEV market growth and innovation. Governments and industry players must invest in infrastructure and entrepreneurship to unlock ASEAN's full potential in the global IEV industry.

References

1. PwC, "Merge Ahead: Electric Vehicles and the Impact on the Automotive Supply Chain," accessed June 1, 2024, https://www.pwc.com/us/en/industries/industrial-products/library/electric-vehicles-supply-chain.html.
2. Mckinsey, "Why the Automotive Future Is Electric," accessed June 1, 2024, https://www.mckinsey.com/industries/automotive-and-assembly/our-insights/why-the-automotive-future-is-electric.
3. Automotive World, "Do Smart EVs Spell the End of the Traditional Supply China," accessed June 1, 2024, https://www.automotiveworld.com/articles/do-smart-evs-spell-the-end-of-the-traditional-supply-chain/.
4. Liu, S., "The Evolution of the Intelligent Electric Vehicle Supply Chain," Forbes, accessed June 1, 2024, https://www.forbes.com/sites/forbestechcouncil/2023/03/01/the-evolution-of-the-intelligent-electric-vehicle-supply-chain/.
5. Liu, S., "China's Bet on Intelligent Electric Vehicles," The Diplomat, accessed June 1, 2024, https://thediplomat.com/2022/12/chinas-bet-on-intelligent-electric-vehicles/.
6. Liu, S., "The Next Frontier of the Global Intelligent Electric Vehicle Industry," Forbes, accessed June 1, 2024, https://www.forbes.com/sites/forbestechcouncil/2022/07/28/the-next-frontier-of-the-global-intelligent-electric-vehicle-industry/.

Chapter 03

The Trickle-Down Effect: Transition to the Autonomy Economy

Economic growth is driven by technological innovation, the most potent factor in the growth equation.

—Robert Solow, Nobel Laureate in Economics

This chapter explores the profound influence of the IEV industry on our societal shift toward the autonomy economy, characterized by the utilization of autonomous machines across various domains, including autonomous vehicles, delivery robots, industrial robots, drones, and home service robots, to deliver goods and services [1]. **The central argument posits that the IEV sector's supply chain, comprising essential elements such as talent, technology, and manufacturing capabilities, will significantly expedite society's transition to the automation age by facilitating the empowerment of diverse segments within the robotics and automation industry** [2]. However, this convergence of IEVs and robotics also poses unique supply chain challenges, particularly concerning resource competition for components like semiconductors and batteries. The chapter commences by elucidating the intertwining supply chains of IEVs and robotics, followed by a case study on China to illustrate the trickle-down effect, ultimately concluding with an exploration of the potential ramifications of the autonomy economy.

3.1. The Overlapping Supply Chains

There is over 80% overlap between the supply chains of IEVs and robotics; hence, increasingly more IEV companies are leveraging this advantage to develop robotics products, as exemplified by Tesla's Optimus, XPeng's humanoid robot, and Xiaomi's CyberOne. This intersection of supply chains is particularly evident in areas such as sensor technology, battery systems, and advanced computing, where both industries demand high-performance components and innovative solutions.

- **Battery and Power Systems:** Both IEVs and robotics rely heavily on advanced battery technology to provide energy-efficient and long-lasting power solutions. The development of lighter, more powerful batteries with faster charging capabilities is crucial for both EVs and a wide range of robotic applications, from industrial automation to consumer robotics.
- **Sensor Technology:** The advancement in sensor technology, crucial for autonomous driving in IEVs, is also a cornerstone in the development of intelligent robotics. Sensors such as LiDAR, radar, and cameras are fundamental in enabling both autonomous vehicles and robots to navigate and interact with their environment safely and efficiently.
- **Artificial Intelligence (AI) and Computing Hardware:** The computing hardware and AI algorithms that power the decision-making processes in IEVs are similarly critical in robotics. The need for real-time processing of vast amounts of data, decision-making capabilities, and learning algorithms is a common requirement, driving innovations in high-performance computing units that are shared across both industries.
- **Software and Connectivity:** The software ecosystems in IEVs, which include systems for vehicle control, data analytics, and user interface, have parallels in robotics. Similarly, connectivity solutions that allow for over-the-air updates and cloud-based services are increasingly important in both sectors, enabling continuous improvement and new functionalities.

Companies that develop versatile, high-quality components that can be used in both IEVs and robotics may find themselves at a competitive advantage. Collaboration and knowledge sharing between the two sectors can accelerate innovation, leading to breakthroughs that benefit both fields.

As technology continues to evolve, the lines between IEVs and robotics will likely become increasingly blurred. The integration of robotic elements in vehicles, such as advanced automation and intelligent assistance systems, and the use of EV technology in robotics signifies a future where the synergy between these two domains becomes more pronounced and mutually beneficial.

The overlapping supply chains of IEVs and robotics represent a dynamic and symbiotic relationship, driving innovation and advancement in both fields. As technology continues to advance, this interplay is likely to become an increasingly important factor in the evolution of both industries.

3.2. Trickle-Down Supply Chain—From IEVs to Robots

As mentioned in the previous chapter, in the past decade, China has poured over $100 billion to boost the growth of the IEV industry. As a result, China now owns the most sophisticated IEV supply chain and is already the largest IEV market in the world. While the dominating argument is that China made this investment to reduce greenhouse gas emissions, I argue that this investment has a much more profound impact on China's economy as the IEV supply chain, including talent pool, technologies, and manufacturing capabilities, will trickle down to empower various segments of the robotics and automation sector, thus accelerating China's economic transformation into the automation age [3].

In the context of globalization, with its enormous supply of low-cost labor, China's economy has benefited from its demographic dividends by serving as the world's factory. However, as labor cost rises and its population ages rapidly, China may no longer enjoy its demographic dividends. In fact, many companies are moving their factories out of China. To avoid falling into the middle-income trap, a structural change to China's economy is imperative.

Historically, only few countries ever succeeded in escaping the middle-income trap, and no country with an aging population ever escaped from the middle-income trap. Nonetheless, traditional economic structural change theories are human-centric, such that the human labor force is the source of productivity. Today, technological innovations enable robots to handle many agricultural, industrial, and service tasks. As robots become more ubiquitous in our daily lives, I believe robots can be the source of productivity and robots can rescue China out of the middle-income trap.

Indeed, we have seen the emergence of various forms of robots integrating into the daily lives of Chinese people, whether it is food delivery robots on the street or cleaning robots at home [4]. For China's economy to transform into the automation age, a complete robotic technology supply chain needs to be developed. The key components of this supply chain include **AI software, batteries, sensors, computer chips, and servo motors. All of these components are required in IEVs as well, albeit in a larger volume and with stricter technical requirements.** China's heavy investment in the IEV industry has generated fierce competition in these technical areas, which will lead to overprovisioning of production capabilities as well as rapid cost reduction. Consequently, the robotics and automation sectors will benefit from this to produce robots that are effective and cost-efficient compared to their human counterparts.

Now let us examine the key components of the robotic technology supply chain. In Figure 3.1, I have compared the robotic technology supply chain market share between China and the United States.

Figure 3.1 Robotic supply chain market share: China vs. the United States.

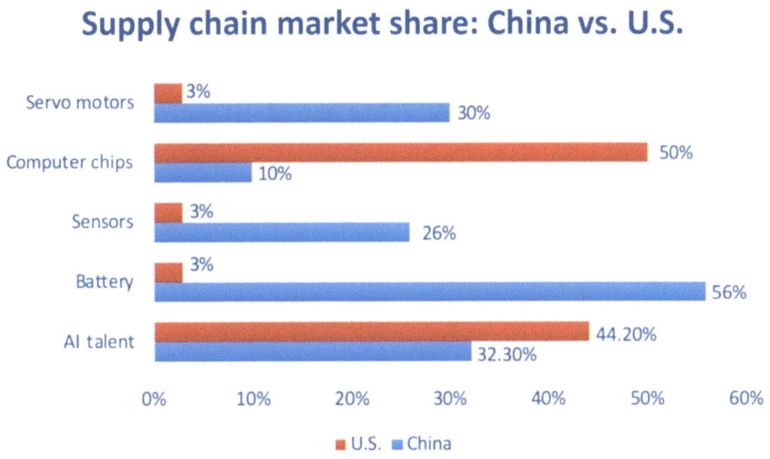

Courtesy of Shaoshan Liu.

- **AI Talents:** Both United States and China are leading in AI talents, with China owning 32.3% and the United States owning 44.2% of the global AI talents. Many AI talents work in IEV companies, such as Tesla in the United States and Baidu in China. From my personal experience, in China, AI talent mobility is very active between the IEV industry and the robotics and automation sector, due to the fact that many AI technologies developed for IEVs can also be used for various types of robots. This is also verified by Tesla's recent move to produce Tesla Bot. A recent trend is that China generates more STEM graduates compared to the United States, and increasingly more technology talents are moving back to China from the United States and Europe. With a thriving IEV industry and a promising automation and robotics sector, China may soon develop the world's biggest AI talent pool.
- **Battery:** As the world's factory in the mobile phone era, many Chinese companies became key iPhone suppliers, especially battery companies. As a result, China is already dominating the global battery market, providing more than 56% of the world's batteries. As a comparison, since battery manufacturing is considered a high-pollution industry, the United States had outsourced most battery manufacturing overseas. Realizing this problem, the US government is trying to bring battery manufacturing back to the United States through the Inflation Reduction Act of 2022.
- **Sensors:** Sensors, such as LiDARs, are essential for IEVs and robots. Although US companies are pioneers in sensor technologies, benefiting from China's booming IEV industry, Chinese sensor suppliers have experienced tremendous growth in recent years. Today, Chinese sensor suppliers own 26% of the global market, whereas US companies only own 3%. In addition, Chinese suppliers are still growing rapidly whereas US companies' market share continues to shrink.
- **Computing Chips:** Although China's booming IEV industry has incentivized the growth of many technical areas, computer chips have always been China's Achilles heel. Today, the United States still dominates the global microprocessor market, especially high-end AI processors. Realizing the strategic importance of microprocessors, the US Congress has recently passed the CHIPS Act of 2022 to further strengthen domestic semiconductor manufacturing, design, and research. In addition, the recent TSMC's opening of a new plant in the United States is another big win for the US semiconductor industry. Without an advanced microprocessor supply chain, China still has to rely on US supply to sustain the growth of its robotics and automation sector in the foreseeable future.
- **Servo Motors:** Servo motors determine the precision and controllability of robots, especially industrial robots. Japanese manufacturers dominate this area due to their technological advantages, but Chinese manufacturers have been growing rapidly due to their cost advantages. Today, Chinese manufacturers have a 30% share of the global market, whereas the US presence is almost negligible. Similar to how China's IEV industry incentivized the growth of battery and sensors supply chain, it is reasonable to project that Chinese servo motor suppliers will grow rapidly with China's robotics and automation sector and they will be able to dominate this market in the foreseeable future.

In summary, China's enormous investment in the IEV industry will not only make China greener, but more importantly accelerate China's economic transformation into the automation age. Examining the robotic technology supply chains reveals that China has gained, or is on the way to gain, dominance in many key technological areas of the robotics and automation sector. If this trend continues, we may soon find ourselves in a world powered by Chinese robots.

3.3. Transition to the Autonomy Economy

Following the previous section, we delve into China's transition to the autonomy economy in this section. The 2023 World Robot Conference took place in Beijing, China, and many companies demonstrated their latest autonomous machine products, from agricultural robots for cherry picking to humanoid robots impersonating ancient Chinese poets, such as Li Bai. Compared to a few years ago, when the autonomous machine sector was dominated by commodity robot vacuums and industrial robots, today autonomous machines have permeated into many aspects of our daily lives, especially in China.

This makes me wonder whether our society is transitioning from the digital economy to the autonomy economy, and if so, what will catalyze the explosive growth of the autonomy economy in the next decade. The past three decades of the world's economic development have been driven by the digital economy and the use of information technology to create, market, distribute, and consume goods and services.

Take the Internet industry, for instance. It has been reported that the Internet industry has accounted for 21% of the GDP growth in mature economies from 2005 to 2010. Specifically, in 2019 the Internet industry contributed $2.1 trillion to the US economy, about 10% of the US GDP, which represents the fourth largest industry of the US economy, only behind real estate, government, and manufacturing. Along with its contribution to the economy, the Internet industry has created nearly six million direct jobs, which accounts for 4% of US employment. In part, the United States has remained a world-leading economy in the past three decades because of its pioneering position in digital technologies, represented by tech conglomerates such as Microsoft, Google, Amazon, Facebook, Intel, and NVIDIA.

With an explosive growth of autonomous machines in both form factors and functions, we may have entered a new technological era, the autonomy economy: the use of autonomous machines, such as autonomous vehicles, delivery robots, industrial robots, drones, and home service robots to provide goods and services. The autonomy economy's impact will be even greater than the digital economy. For instance, the US transportation sector is worth $1.9 trillion, or 8.4% of the US GDP. The ubiquitous deployment of one form of autonomous machines, autonomous vehicles, will completely transform the transportation sector.

Similarly, various forms of autonomous machines will transform almost every existing industry through replacing human workers or service providers. Hence, a **natural deduction is that the countries that dominate autonomous machine technologies will dominate the world's economy in the coming decades**.

The transition to the autonomy economy seems to be happening in China first, as various forms of autonomous machines have gradually become an integral part of people's daily lives. When you check into a hotel, it is very likely that a delivery robot is going to bring what you need to your room. When you order food and grocery delivery, it is robots that deliver the food from the restaurant or grocery from the supermarket to your doorstep. At night, it is likely that you will witness a fabulous air show performed by a swarm of robotic drones. At home, robotic vacuum cleaners have become standard home appliances for many Chinese families, just like refrigerators and air conditioners.

However, a hidden caveat behind this superficial prosperity of the autonomous machine sector is the scalability of autonomous machine technologies. The scalability of digital technologies enabled the digital economy to reach economies of scale and transformed the digital economy into the world's economic growth engine. For instance, Google employs less than 30,000 engineers while serving over 4.3 billion users. The scalability of digital economy companies is mainly constrained by available compute resources and data instead of by the number of engineers. As a result, these companies are able to provide their service at mass scale at a very low price.

In contrast, although many autonomous machine products have emerged in the past decades, autonomous machine technologies are not yet able to achieve economies of scale. An enormous amount of engineering effort is required to tailor-make autonomous machines for a specific function in a specific environment. When the operating environment or the function of the autonomous machines changes, more engineering efforts are demanded. As a result, the costs of deploying autonomous machines to fulfill a specific task are often higher compared to human labor. This explains the emergence of various form factors of autonomous machines in the past decade, but not ubiquitous deployments of these autonomous machines in our daily lives yet [5].

Hence, **a further deduction is that the countries which develop scalable autonomous machine technologies will dominate the world's economy in the coming decades**. As examined in the previous section, China currently leads in the hardware supply chain of autonomous machines and possesses significant advantages in batteries, servo motors, sensors, and manufacturing capabilities. These are key areas for achieving economies of scale for autonomous machine manufacturing.

On the other hand, the United States leads in autonomous software technologies and computing semiconductors. Particularly, the software technology gap has been amplified by the recent emergence of large language models (LLMs) in the United States. The application of LLMs demonstrates that AI can be general purpose instead of tailored for a specific application. The variations of LLM technologies can be infused into autonomous machines to perform various tasks, thus achieving scalability.

The technologies in China and the United States perfectly complement each other; however, the current US–China tech tension puts the future of the autonomy economy at a crossroads. If these two countries can still cooperate, then we may see much faster progress toward achieving economies of scale in the autonomy economy.

In contrast, if both countries decide to go their own ways, China has to invest heavily in software and semiconductors in order to catch up with the United States, whereas the United States has to reshape its global hardware supply chain in order to supply affordable autonomous machine products. The competition between these two countries in tech will not only incur very high costs to duplicate autonomy economy systems, but also delay the integration of scalable autonomy economy into society as it takes years to make up for the missing part in each system.

The intertwining of the IEV industry with the robotics sector is driving profound changes in our society. This chapter has illustrated the extent of their overlapping supply chains, shedding light on how shared components and technologies are propelling innovation in both fields. Furthermore, it has utilized China as a case study to illustrate the pivotal role of IEV in the transition to the autonomy economy. China is not only a dominant force in IEVs but also a rising power in the automation and robotics sector.

3.4. Summary

In conclusion, as we witness the emergence of the autonomy economy, characterized by the pervasive presence of autonomous machines, we find ourselves on the cusp of a new technological era with far-reaching economic implications. However, the path forward is not without its challenges, particularly amid US–China tech tensions. The future of the autonomy economy hinges on international collaboration and innovation, as well as the development of scalable technologies. Whether it is through cooperation or competition, the choices made by nations and industries will significantly shape the trajectory of this evolving landscape. One thing is certain: the autonomy economy is upon us, and its impact on our world is destined to be profound and transformative.

References

1. Liu, S., "Shaping the Outlook for the Autonomy Economy," *Communications of the ACM* 67, no. 6 (2024): 10-12.
2. Liu, S., "China's Trickle-Down Supply Chain from Intelligent Electric Vehicles," The Diplomat, accessed June 1, 2024, https://thediplomat.com/2022/12/chinas-trickle-down-supply-chain-from-intelligent-electric-vehicles-to-robots/.
3. Liu, S., "China's Rise as a Robotic Nation," Forbes, accessed June 1, 2024, https://www.forbes.com/sites/forbestechcouncil/2022/04/29/chinas-rise-as-a-robotic-nation/.
4. Liu, S., "The Transition to the Autonomy Economy and China-U.S. Tech Competition," The Diplomat, accessed June 1, 2024, https://thediplomat.com/2023/08/the-transition-to-the-autonomy-economy-and-china-us-tech-competition/.
5. Wu, S., Yu, B., Liu, S. and Zhu, Y., "Autonomy 2.0: The Quest for Economies of Scale," arXiv preprint arXiv:2307.03973, 2023, accessed September 4, 2024, https://arxiv.org/abs/2307.03973.

Chapter 04

Scenario-Driven Car Making

The car is the closest thing we will ever create to something that is alive.

—Sir William Lyons, Founder of Jaguar

In this chapter, I introduce "scenario-driven car making" as the new approach for IEVs. I envision IEVs evolving similarly to smartphones, with diverse usage scenarios beyond mobility. I argue that several IEV companies, such as Li Auto, Canoo, BeyonCa, and HiPhi, have already embraced this concept, tailoring their vehicles for specific scenarios. This shift requires an innovative top-down approach, where usage scenarios define vehicle designs, enabled by modular vehicle engineering, akin to assembling LEGO blocks, reducing development time and costs. This approach signals a shift from a technology-driven bottom-up process to a scenario-driven top-down one, emphasizing the importance of understanding customer needs and developing vehicles accordingly. Two case studies are also presented in this chapter: the first one is an AMC, a medical usage scenario for IEVs, and the second one is infrastructure–vehicle cooperative autonomous driving (IVCAD), a new usage scenario for the road transportation system.

4.1. Building Cars for Usage Scenarios

Coming from a pure technology background, I have spent the last decade developing autonomous driving technologies. More than a year ago, I embarked on an exciting new journey of making IEVs, of which autonomous driving is only one component. After deep diving into the car making process for a while, I have been awed by the complexity of car engineering, challenged by the cultural clash of traditional car making and Internet-style fast-paced innovations, touched by the ever-lasting beauty of car designs, inspired by the wide spectrum of usage scenarios of IEVs.

At the very beginning of 2023, I had an epiphany, realizing that we have come to the age of scenario-driven car making, such that IEVs should serve a very clear and detailed customer need. With the sole mission of fulfilling this need, we should then gradually unfold the complexity of car making. I call this new process scenario-driven car making [1].

In this section, I attempt to explain the concept of scenario-driven car making. The IEV industry is evolving through three stages of development: electrification, intelligence, and ecosystem. After more than a decade of development, various IEV companies, such as Tesla, NIO, and XPeng, have laid a very good foundation on electrification and intelligence, and it will be extremely challenging for latecomers to challenge their established positions in electrification and intelligence. Thus, it will be a smart move for any new IEV company to rely on the existing supply chain for electrification and intelligence but focus on proprietary technologies to enable a few usage scenarios.

In a way, an IEV is similar to a smartphone, such that an application ecosystem can be developed for IEVs. Some people use their phones mainly for work purposes, some use their phones mainly for entertainment, and some use their phones as health monitors. Similarly, I project that the IEV industry will evolve similarly, such that with an IEV ecosystem developed, many usage scenarios beyond mobility will emerge.

Indeed, I have already observed some examples of scenario-driven car making in the IEV industry. China's Li Auto has developed a series of "soccer mom" cars, to meet the needs of picking up the children, taking a family trip, and other family-oriented usage scenarios. The key features of Li Auto's IEVs include a spacious interior environment and large screens for entertainment. Canoo, a US-based IEV company, develops "lifestyle" vehicles, specifically targeting camping usage scenarios. BeyonCa, a newcomer in the IEV space, focuses on building luxury vehicles targeting in-car health monitoring. HiPhi, another Chinese IEV brand, develops robot cars for customers to program and interact with, such that a user can make the car dance or play a musical show.

As more and more scenario-driven IEVs come to market, we need to develop an innovative car making process to develop vehicles for specific usage scenarios, instead of infusing new usage scenarios on existing platforms. The traditional car making process is a technology-driven bottom-up process, such that automotive engineers define the key parameters of the chassis platform, including dimensions, mass, torque, gear ratios, aerodynamic drag forces, etc. Then, based on this platform, designers derive various models for different customer groups. On top of these models, then product managers elaborate on usage scenarios.

Scenario-driven car making, in contrast, is a top-down process such that the usage scenarios are defined first, and then the car's exterior and interior designs are defined surrounding the usage scenarios. Finally, different technologies, such as in-cabin intelligence, autonomous driving, and chassis, are scouted from the technology supply chain to empower the usage scenarios. What makes the top-down scenarios-driven car making process possible is the emergence of modular vehicle engineering, exemplified by technologies such as skateboard chassis and centralized electrical and electronic (E/E) architectures.

These standardized modules greatly reduce the cost and development time of vehicle engineering. More importantly, these standardized modules provide well-defined interfaces and abstractions for IEV OEMs to build cars for specific usage scenarios, similar to how our kids can unlock their imagination using LEGO building blocks.

Case Study: Autonomous Mobile Clinics

AMCs are autonomous vehicles equipped with medical diagnostic equipment, AI doctor software, and telemedicine capability. AMCs have the potential to revolutionize healthcare delivery by bringing healthcare services to the patient at the touch of a button [2, 3]. Specifically, AMCs consist of two parts: the "autonomous" part allows AMCs to autonomously navigate to places where healthcare services are most needed, thus solving the mobility problem. The "mobile clinic" part provides on-site healthcare access, thus solving the access problem. In addition, the AI doctor within AMCs can serve unlimited patients, thus potentially eliminating the cost barrier. This section uses AMC as an example to explain the scenario-driven car making process [4, 5].

Usage Scenario: Affordable Anywhere Anytime Healthcare Access

To create a more efficient healthcare system, we can start by enabling affordable anywhere anytime healthcare access through technological innovations, starting with primary care. Three technical challenges immediately present themselves. The first challenge is access itself; people are denied access to healthcare services for various reasons, including lack of mobility, particularly for disabled people and the elderly, and unpleasant user experiences, especially for countries with overloaded public healthcare systems. An AMC that comes to you at the touch of a button solves the access problem.

The second challenge is equity; while mobility enables access, it does not guarantee equal access. The reality is that healthcare resources are unevenly distributed, and this distribution is often strongly correlated with wealth distribution. The equity problem again can be solved by technology; telemedicine on the mobile clinic connects the patient to medical experts worldwide, for the first time in human history, disrupting the uneven distribution barrier to truly enable equal healthcare access.

The third challenge is efficiency; as the overall healthcare cost is climbing, people have high hopes for new technologies to improve healthcare access and quality while minimizing costs. We can consolidate clean health monitoring and diagnostic data through AMCs to train an AI doctor for the first time, which has been proven to be highly cost-efficient. An AI doctor will handle most prescreening and diagnostics tasks in primary care, freeing human doctors for more sophisticated tasks.

Each AMC is equipped with a nurse to help patients perform basic health screening and connect to remote doctors. Also, the interior design of AMCs is configurable to adapt to the needs of different medical specialities. For instance, it can be configured to a pulmonary care clinic through integrating equipment such as nebulizers and suction machines. However, the enabling technologies remain the same for different usage scenarios, which we will discuss later in this section.

Health Benefit Analysis

After understanding the usage scenario, we now analyze the potential benefits of AMCs. To address the aforementioned challenges, AMCs provide a revolutionary and effective way of healthcare service delivery. The idea of AMCs is not new. Concept products such as Toyota e-Care have emerged in recent years, but only stayed in the concept stage.

As shown in **Figure 4.1**, to enable a universal AMC network, a three-stage technical roadmap needs to be achieved. In stage 1, we focus on solving the inequity challenge in the existing healthcare system by combining autonomous mobility, telemedicine, and primary care. In stage 2, we develop an AI doctor for primary care, an AI doctor that we foster from infancy to adulthood with clean healthcare data. With the AI primary care doctor, we can solve the inefficiency problem. In stage 3, after we have proven that the AMC network can truly solve the primary care problem, we shall open up the platform for all medical verticals, thus enabling universal healthcare through this whole new system.

Figure 4.1 A three-stage AMC technical roadmap.

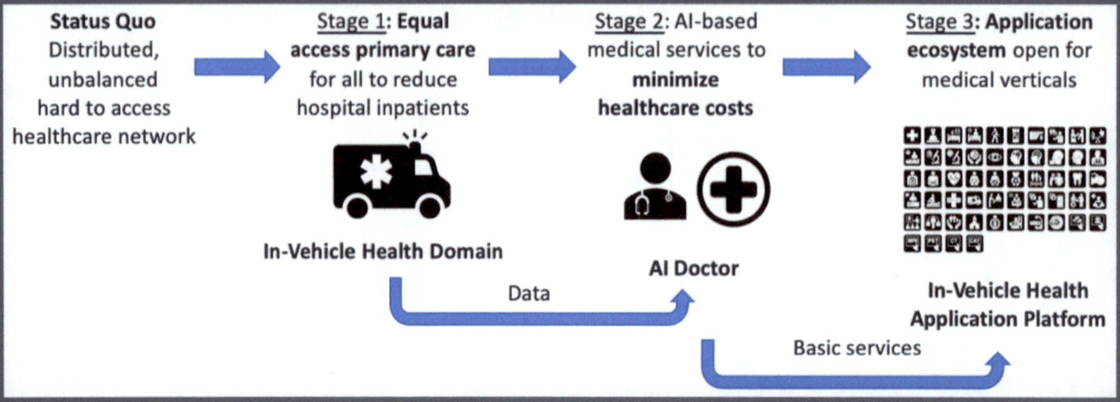

AMCs also bring along multiple immediate clinical benefits. First, AMCs provide a perfect environment for a point-of-care test (POCT), which refers to multiple clinical tests carried out at the point of care near the patients, rather than from a clinical laboratory. When dealing with patients with infectious diseases, lab tests such as white blood cell test and C-reactive protein are of the utmost importance. AMCs provide an isolated environment for POCT at the location where the patients reside, effectively constraining the spread of infectious diseases.

Under a pandemic situation such as COVID-19, medical resources become even more scarce and many patients suffering from serious health conditions, such as myocardial infarction, stroke, or diabetes, fail to get proper testing, such as creatine kinase myocardial band (CK-MB) and blood glucose, let alone treatments. The multiplex POCT model includes regular gases of blood, metabolites, as well as patient's electrolytes, and various immunoassays.

By following the multiplex POCT model to integrate medical instruments in AMCs, we can greatly reduce the lag time on transferring blood samples to the laboratory as well as enhance the diagnostic efficiency. Taking acute myocardial blood marker test as an example, it has been confirmed that POCT introduces a mean reduction of 110 min of test result turnaround time compared with traditional laboratory testing methods; the same benefit has been observed in the emergency department (ED) as well.

The second benefit of AMCs is the combination of the multiplex POCT model with telemedicine to carry out online healthcare consultation services to patients. With this capability, remote doctors can review patient's lab test results in real time; hence, the doctors are able to perform consecutive screening and monitoring, as well as to provide simple instructions on the treatment process. These all take place at the point of care without having physical contact with the patients, greatly reducing the risks of contracting infectious diseases.

In the *New England Journal of Medicine* report "Telemedicine Is Mainstream Care Delivery," the authors conclude that "Mobile + Medical care" is an essential way to improve healthcare efficiency. The report strongly advocates major hospitals to develop remote consultation systems, whether it is mobile or static, to address the imminent problem of healthcare access [6].

AMCs address this exact problem, and now we summarize the potential benefits AMCs bring to healthcare management:

1. **Optimal medical resources allocation:** The AMC remote consultation system can break geographical boundaries, not only enabling patients in different areas to access medical services, but also improving the level of medical services in poor population, so that medical resources can be allocated and used more reasonably. The ultimate goal of the AMC project is to reach universal healthcare access. Unfortunately, today half the world's population lacks access to essential health services. The per capita healthcare spending in the United States is $12,000, whereas the per capita healthcare spending in the LDCs is merely $42. AMCs have the potential to narrow this tremendous resource allocation gap and enable a healthier population in LDCs, which is essential to boost the economy in LDCs. We are currently conducting a thorough study to understand the full impact of AMCs on the world economy.
2. **Affordable medical training:** One of the key roadblocks for universal healthcare access is the lack of healthcare professionals. Today, the global medical education market size is only around USD 30 Billion, and the average cost of training a medical doctor in the United States is around $200,000. This is an extremely high cost barrier for people in LDCs. AMCs can not only be used for disease screening, diagnosis, and treatment, but also for telemedicine training, education, and international telemedicine exchanges. From this perspective, with AMCs, hospitals across the world can form medical education alliances to develop medical training curricula and to deliver medical training to medical trainees and healthcare professionals around the world through AMCs. Improving medical skills and leveling the healthcare service qualities across the globe is another key objective of the AMC project.

3. **Healthcare cost reduction:** On average, a primary care doctor can treat at most 20 patients per day, and the average annual salary of a primary care patient is around $200,000 in the United States This cost is unbearable for people in LDCs. In contrast, AI doctors within AMCs can treat unlimited patients with virtually no additional cost. Therefore, the AMC platform is able to minimize healthcare costs. Today, many AI doctors are already able to perform basic disease screening to increase the number of patients a primary care doctor can treat today. As a next step of the development of AMC, we plan to focus on developing AI doctors for the top ten diseases in LDCs.

Enabling Technologies

After defining AMC's usage scenarios and benefits, we now review the enabling technologies necessary to make this product feasible. We briefly introduce the enabling technologies for AMCs. AMCs rely on multiple pillars of technical areas, including autonomous driving, telemedicine, medical equipment integration, and electronic health records:

- **Autonomous Driving:** Level 4 autonomous driving is required for autonomous healthcare delivery. Each AMC needs to consist of a fully drive-by-wire chassis with fail-over backup systems, essential sensors for panoramic perception, a reliable localization system that can achieve decimeter level accuracy, a planning and control system that enables the vehicle to autonomously navigate through the target environment, as well as a high-definition (HD) map of the operation area. This technical area will be further discussed in Chapter 7.

- **Telemedicine:** Network connectivity is key to guaranteeing the quality-of-service (QoS) of telemedicine applications. There are two categories of data communication for in-vehicle healthcare services: healthcare sensing data and video data stream. Healthcare sensing data involve the transmission of raw sensing data collected from healthcare sensors, such as imaging devices, whereas video streams allow interactive communication between the patient and the remote doctor. Connectivity for IEVs will be further discussed in Chapter 11.

- **Medical Equipment Integration:** A mobile clinic requires the integration of many medical equipment. However, each medical equipment manufacturer enforces its own data standard, preventing the seamless healthcare delivery experience, especially that the equipment deployment needs to be highly customized to the specific healthcare service scenarios. A standard method of integrating and making sense of the enormous amount of sensing data is required. A detailed technical discussion on this topic will be discussed in Chapter 9.

- **Electronic Health Records:** Electronic health records are key to enabling integrated care and to the training of AI doctors. With AMCs, for the first time we are able to collect integrated health data instead of having to collect fragmented health data from each department. This is the prerequisite for AMCs serving as a touch point of integrated care. In addition, by combining the integrated health data and diagnostics from the remote doctor, we can develop AI systems to handle basic diagnostic functions to reduce overall healthcare costs.

- **AI Doctors:** Health AI engines incorporated into AMCs can serve an unlimited number of patients, and only cases that cannot be handled by the software would require the intervention of remote doctors, thus potentially reducing the cost for the human workforce. While AMCs serve as mobility platforms for essential health services provisioning, the health AI engines incorporated into AMCs determine the reach and coverage of AMCs. In essence, health AI engines are applications on AMCs, similar to how various applications expand the usage scenarios of smartphones. This way, AMCs should be configured with different medical devices and health AI engines as the deployment situation demands [7].

This section delves into a comprehensive case study that focuses on the development of AMCs through a structured three-step approach. Initially, the usage scenarios for AMCs are meticulously defined, laying the groundwork for their practical application. Subsequently, a thorough health benefit analysis is conducted, shedding light on the profound impact and significance of AMCs in the realm of healthcare. Finally, the section concludes by identifying the key enabling technologies that are crucial for realizing the feasibility of AMCs. It is important to note that AMCs represent the ultimate integration of healthcare usage scenarios into the car making process, with an emerging trend of IEV OEMs incorporating healthcare monitoring into their smart cockpit designs.

Case Study: IVCAD

IVCAD is the utilization of infrastructure-side sensing and computing capabilities to empower autonomous vehicles to be smarter and to make the whole transportation system more efficient. IVCAD is a rising technology area under immense debate, with most of the criticism around the economic rationality of infrastructure investments. In this section, we identify the usage scenarios for IVCAD, analyze the business case for IVCAD, and explain the enabling technologies to make IVCAD feasible [8].

Usage Scenario: Ubiquitous and Safer Autonomous Driving

Autonomous driving is a thriving sector with the promise to bring enormous benefits to our society. In the past few years, we have observed tremendous progress in the technical development of autonomous driving technologies. However, the key question is: **When will the technology be fully ready for us to realize all the benefits promised?** Many who work in the field remain quite pessimistic regarding this question and argue that autonomous driving technologies will be ready only after 2040, still about two decades away [9].

The pessimism is regarding the traditional approach of autonomous driving, which only utilizes the limited on-vehicle intelligence, such as sensing and computing capabilities. With limited sensing and computing capabilities, on-vehicle sensors can fail in many situations: objects out of the sensing ranges cannot be detected, cameras are challenged by varying lighting conditions, and LiDARs can be confused by varying weather conditions such as fog and snowfall. Unusual, unfamiliar, and unstructured situations, such as accidents, road work, or a fast-approaching emergency response vehicle, can also be hard to classify.

The key problem is that it is challenging for autonomous vehicles to gather comprehensive information from their environment, and they are not smart enough to operate on their own with only limited information. One promising solution to this problem is IVCAD, or **the utilization of infrastructure-side sensing and computing capabilities to empower autonomous vehicles to be smarter and to make the whole transportation system more efficient** [10, 11].

Business Analysis

According to the World Economic Forum, the intelligent vehicle industry is estimated to create more than $67 billion in economic value and more than $3.1 trillion in social benefits globally [12]. The market potential for autonomous vehicles in the coming decades is enormous: by 2035, China alone will deploy about 8.6 million self-driving cars, with about 3.4 million of them having fully autonomous driving capabilities and 5.2 million having semiautonomous driving capabilities [13]. Furthermore, it has been predicted by multiple sources that it will take 15–20 years for the global market share of self-driving vehicles to reach 25%. The impact of autonomous driving technology on our economy is unprecedented, particularly in the following areas.

Traffic Safety Enhancement: Traffic accident fatalities are the biggest problem facing the transportation industry worldwide. In the United States alone, it is estimated that at least 35,000 people are killed in car accidents each year, and the number in China is about 260,000 [14]. According to the World Health Organization, nearly 1.3 million people worldwide are killed in highway accidents each year [15]. In the United States alone, traffic injuries and deaths lead to at least $625 billion in economic losses each year [16]. The universal deployment of autonomous driving technology will drop this number to close to zero [17].

Traffic Congestion Reduction: Traffic congestion is a core traffic problem faced by almost every major city in the world. In the United States, traffic congestion causes an annual economic loss of $121 billion [18]. In addition, it is estimated that 23–45% of urban traffic congestion occurs at road crossings. Traffic lights and stop signs do not work because they are stationary and cannot take traffic flow into account. Green or red lights are set at fixed intervals in advance, regardless of the volume of traffic in a given direction. Once autonomous vehicles become ubiquitous, these vehicles will interact with intelligent traffic systems to optimize the flow of traffic at road intersections, and the interval between traffic lights can be dynamically adjusted to the flow of traffic. This can ease congestion by greatly improving the efficiency of vehicle traffic.

Air Pollution Reduction: Cars are one of the leading causes of air pollution. Studies have shown that autonomous driving technology can improve energy efficiency by 4–10% over manual driving through smoother acceleration and deceleration, and cars cause 40% more pollution when waiting at red lights or in traffic jams compared to when the vehicle is moving [19]. By universally adopting autonomous vehicles, emissions can be reduced by more than 10% [20]. It is estimated that carbon pollution from transportation in the United States costs over $200 billion [21].

Figure 4.2 summarizes the US economic impacts that can be improved by autonomous driving technologies, an astonishing $946 billion, in which traffic safety occupies 66%, air pollution occupies 21%, and traffic congestion occupies 13%. Hence, any improvement brought by autonomous driving technologies in these areas will make a huge impact on the US economy.

How much safer is IVCAD compared to on-vehicle-only autonomous driving? As for the cause of traffic accidents, it has been observed that human drivers have a crash rate of 4.2 accidents per million miles (PMM), while the current autonomous vehicle crash rate is 3.2 crashes PMM [16]. This means that, on average, autonomous vehicles deliver a significant safety improvement compared to their human counterparts. However, there are corner cases, such as blind spots, which neither human drivers nor autonomous vehicles can handle without the help of an intelligent infrastructure.

Figure 4.2 US economic impacts that can be optimized by autonomous driving technologies.

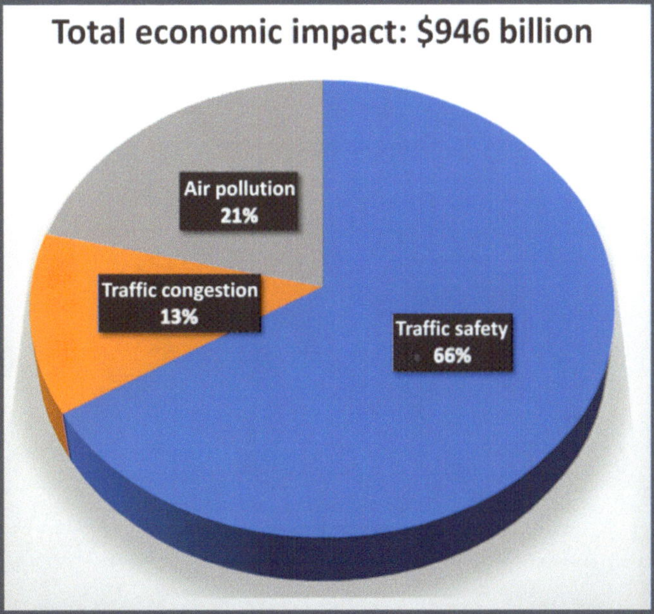

Courtesy of Shaoshan Liu.

Indeed, IVCAD provides a comprehensive global view of the traffic conditions, thus not only eliminating blind spots but also greatly extending the autonomous vehicles' perception distance so as to provide more reaction time for the vehicles [22]. With blind-spot reduction and extended perception afforded by IVCAD, we have observed in our deployments that the autonomous vehicle disengagement rate (the frequency for a human driver to take over control of the autonomous vehicle due to situations that cannot be handled by the autonomous vehicle) has dropped by more than 90%; thus, IVCAD should further drop the autonomous vehicle accident ratio by 90%, from 3.2 crashes PMM to 0.32 crashes PMM and eventually drops to zero as IVCAD continuously improves. This implies the universal deployment of IVCAD in the United States alone could lead to a reduction in traffic safety economic impact by a whopping $578 billion.

How much more efficient is IVCAD compared to on-vehicle-only autonomous driving? The cost of traffic congestion alone is estimated to be greater than $200 billion in four Western countries: France, Germany, the United Kingdom, and the United States. For an average US citizen, congestion costs 99 h of their time and USD 1377 each year. The unique advantage of IVCAD is its capability to generate comprehensive traffic information and adjust traffic accordingly to achieve maximum efficiency. For instance, IVCAD can track vehicles and adjust traffic lights when there are fewer or no cars approaching, and even maximize traffic flow through global traffic planning. At our deployment site, we verified this finding by utilizing per-vehicle tracking data generated and adjusting traffic lights accordingly utilizing IVCAD; we managed to improve traffic efficiency by 40%. This is simply a first step, as there is still plenty of room for improvement. This implies the universal deployment of IVCAD in the United States alone could lead to a reduction in air pollution and traffic congestion economic impact by at least $128 billion.

What is the turning point for IVCAD to make economic sense from the pure hardware deployment cost perspective? There are an estimated 1.2 billion vehicles on the world's roads today, and this is expected to grow to two billion by 2035. On the other hand, in the United States the road capacity is designed to process 1900 vehicles per hour per lane, and there are only about four million miles of public roads in the United States. Converting all the cars in the world into autonomous vehicles would mean an enormous social cost, as autonomous vehicles are still quite expensive to build. A better approach would be to invest in the road infrastructure and make it "intelligent." Gradually, as IVCAD becomes more powerful, we can then migrate more of the computation workload from the vehicles to the roads, hence greatly reducing the hardware and energy costs of autonomous vehicle deployments. Based on our deployment data, it costs at least $100,000 to convert a vehicle into a robotaxi. Considering economies of scale, we project the cost to be around $30,000 for universal deployment. Also, the mass deployment of IVCAD would cost around $100,000 per mile, and only a simple receiver on the vehicle is needed for receiving commands from IVCAD. There are about 300 million registered vehicles in the United States and 4 million miles of public road. We further assume that due to the efficiency increase, only 20% of the vehicles are needed, which drops the required number of vehicles to 60 million.

In Figure 4.3 we show the cost–benefit analysis of IVCAD for the United States. The results show converting all vehicles to robotaxi would increase the total economic impact from $946 billion to $2597 billion, which is against the very purpose of autonomous driving and in a way proving that maybe robotaxi can only serve the select few who can afford it. On the other hand, converting the road infrastructure to IVCAD would bring down the total economic impact from $946 billion to $640 billion, a 32% saving on the impact to the US economy alone.

Figure 4.3 Cost–benefit analysis of IVCAD.

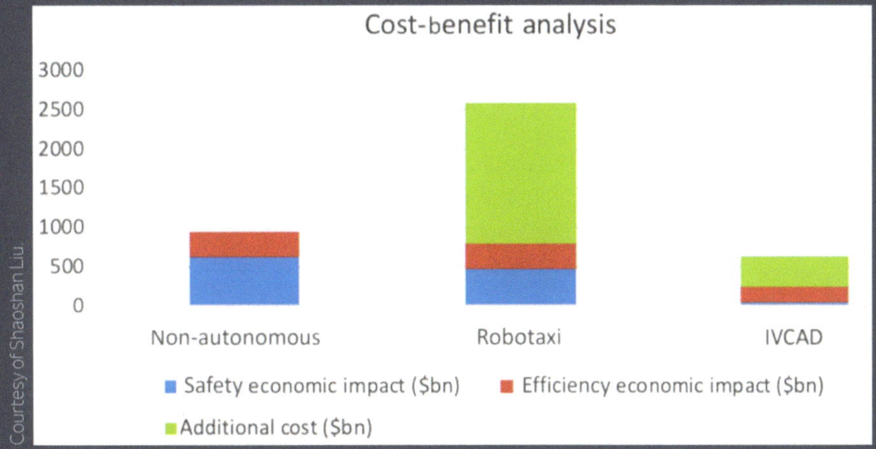

Enabling Technologies

IVCAD system consists of systems on vehicles (SoVs), systems on roads (SoRs), the intelligent transportation cloud system (ITCS), the engineering system, and the control center [11]. The SoRs provide local perception results to the SoVs for blind spot elimination and extended perception to improve safety. Meanwhile, the SoRs process incoming sensor data and send the extracted semantic data to the ITCS for further processing. The ITCS fuses all incoming semantic data to generate global perception and planning information, and then the control center can dispatch global traffic information, navigation plans, and control commands to the SoVs to achieve optimal traffic efficiency. The engineering system is simulation-centric, in which autonomous driving software is developed and verified in a virtual environment before physical deployments. The simulation-centric system brings advantages that maximize development productivity and minimize development and testing costs, but also challenges on how to virtually represent a realistic physical world.

IVCAD's development follows the three stages listed below:

- **Stage 1:** Infrastructure-augmented autonomous driving (IAAD): In this stage, autonomous vehicles fuse both vehicle-side and infrastructure-side perception outputs to improve the safety of autonomous driving. In this stage, both the SoVs and the SoRs are equipped with full sensing and computing capability, and we aim to fuse the information from both the SoRs and the SoVs to improve safety and to achieve maximum computing and energy efficiency.
- **Stage 2:** Infrastructure-guided autonomous driving (IGAD): In this stage, autonomous vehicles can offload all the proactive perception tasks to the infrastructure to reduce per-vehicle deployment costs. In this stage, while the SoRs are equipped with full sensing and computing capability, we can greatly reduce the on-vehicle hardware deployment (both computing and sensing) costs. We aim to design secure, reliable, and safe infrastructure supports for autonomous vehicles.

- **Stage 3:** Infrastructure-planned autonomous driving (IPAD): In this stage, the infrastructure takes care of both perception and planning, thus achieving maximum traffic efficiency and cost efficiency. In this stage, the SoVs are equipped with only basic sensing and computing capabilities, and we aim to optimize global traffic efficiency by centrally planning the trajectory of each vehicle.

Indeed, while IVCAD holds immense potential, it is not without its challenges and hurdles. The need for substantial infrastructure investments, coupled with the requirement for all IEVs to be equipped with additional hardware, poses significant obstacles. Achieving consensus among IEV OEMs and establishing standardization for IVCAD technologies are complex tasks that demand unified industry efforts. Moreover, concerns related to safety and user privacy remain paramount, necessitating robust regulatory frameworks and comprehensive safeguards. The successful realization of IVCAD's business case hinges on a concerted, centralized effort, potentially led by government authorities, to drive widespread IVCAD adoption, addressing these formidable challenges in the process.

4.2. Summary

In conclusion, this chapter highlights the transition from a traditional technology-driven approach to a scenario-driven method in car manufacturing, emphasizing the significance of aligning vehicle development with customer needs. The process begins by comprehensively understanding the specific usage scenario for each IEV product. Subsequently, a thorough analysis of the business case for the product is conducted to determine its feasibility. Finally, we delve into the enabling technologies required for the product's realization, using modular vehicle engineering principles that simplify development, akin to assembling LEGO blocks. This approach is designed to streamline development timelines and reduce costs while prioritizing the customization of IEVs to meet distinct usage scenarios.

References

1. Liu, S., "An Introduction to Scenario-Driven Car Making," Forbes, accessed June 1, 2024, https://www.forbes.com/sites/forbestechcouncil/2023/01/27/an-introduction-to-scenario-driven-car-making/.
2. Liu, S., "From Barefoot Doctors to Autonomous Mobile Clinics," Forbes, accessed June 1, 2024, https://www.forbes.com/sites/forbestechcouncil/2022/06/27/from-barefoot-doctors-to-autonomous-mobile-clinics/.
3. Liu, S. and Kong, A., "How Autonomous Mobile Clinics Can Transform Healthcare in Least Developed Countries," World Economic Forum, accessed June 1, 2024, https://www.weforum.org/agenda/2022/07/autonomous-mobile-clinics-healthcare/.
4. Liu, S., Kong, A., Huang, Y., and Liu, X., "Autonomous Mobile Clinics," *Bulletin of the World Health Organization* 100, no. 9 (2022): 527.
5. Liu, S., Huang, Y., and Shi, L., "Autonomous Mobile Clinics: Empowering Affordable Anywhere, Anytime Healthcare Access," *IEEE Engineering Management Review* 50, no. 4 (2022): 147-154.

6. Hollander, J. and Sharma, R., "Telemedicine Is Mainstream Care Delivery," *NEJM Catalyst Innovations in Care Delivery* 3, no. 7 (2022), doi:https://doi.org/10.1056/CAT.22.0176.

7. Yang, T.T., Yang, T.T., Liu, A., Tang, J. et al., "AICOM-MP: An AI-Based Monkeypox Detector for Resource-Constrained Environments," arXiv preprint arXiv: 2211.14313, 2022.

8. Liu, S., "The Business Case for Infrastructure-Vehicle Cooperative Autonomous Driving," *IEEE Engineering Management Review* 50, no. 2 (2022): 189-194.

9. Liu, S., Peng, J., and Gaudiot, J.L., "Computer, Drive My Car!" *Computer* 50, no. 1 (2017): 8.

10. Oliver, N., Potocnik, K. and Calvard, T., "To Make Self-Driving Cars Safe, We Also Need Better Roads and Infrastructure," *Harvard Business Review*, 2018.

11. Liu, S., Yu, B., Tang, J. and Zhu, Q., "Towards Fully Intelligent Transportation through Infrastructure-Vehicle Cooperative Autonomous Driving: Challenges and Opportunities," in *2021 58th ACM/IEEE Design Automation Conference (DAC)*, San Francisco, 2021, 1323-1326, IEEE.

12. Weindelt, B., "Digital Transformation of Industries: Automotive Industry," World Economic Forum in Collaboration with Accenture (Vol. 4), 2016.

13. China Daily, "Officials Want to Open Way for Autonomous Driving," 2016.

14. New York Times, "Beijing's Electric Bikes, the Wheels of E-Commerce, Face Traffic Backlash," 2016.

15. World Health Organization, "Global Health Observatory Data: Number of Road Traffic Deaths," 2010.

16. Anderson, J.M., Nidhi, K., Stanley, K.D., Sorensen, P. et al., *Autonomous Vehicle Technology: A Guide for Policymakers* (Santa Monica: Rand Corporation, 2014).

17. Liu, S. and Gaudiot, J.L., "Autonomous Vehicles Lite Self-Driving Technologies Should Start Small, Go Slow," *IEEE Spectrum* 57, no. 3 (2020): 36-49.

18. Mendez, V.M., Monje, C.A. and White, V., "Beyond Traffic: Trends and Choices 2045—A National Dialogue about Future Transportation Opportunities and Challenges," in: Meyer, G. and Shaheen, S. (eds), *Disrupting Mobility* (Cham: Springer, 2017), 3-20.

19. Schlossberg, T., "Stuck in Traffic, Polluting the inside of Our Cars," *New York Times*, 2016.

20. Fagnant, D.J. and Kockelman, K.M., "The Travel and Environmental Implications of Shared Autonomous Vehicles, Using Agent-Based Model Scenarios," *Transportation Research Part C: Emerging Technologies* 40 (2014): 1-13.

21. EPA, "Carbon Pollution from Transportation," accessed June 1, 2024, https://www.epa.gov/transportation-air-pollution-and-climate-change/carbon-pollution-transportation.

22. Zhang, K., Liu, Y., Zhang, W., and Liu, S., "π-Learner: A Lifelong Roadside Learning Framework for Infrastructure Augmented Autonomous Driving," *Computer* 55 (2022): 30-39.

Chapter | 05

Project Management in the Making of Intelligent Electric Vehicles

Project management in car manufacturing is like conducting an orchestra; it requires perfect harmony between precision engineering, innovative design, and meticulous coordination to create a masterpiece on wheels.

—Anonymous

This chapter offers an in-depth exploration of the unique product and project management challenges encountered in the development of IEVs. We begin by analyzing the cultural clashes that emerge in the production of IEVs. Unlike traditional automobile manufacturing, IEVs require a convergence of expertise from diverse fields such as automotive engineering, AI, and software engineering. This necessitates a novel approach to collaboration and problem-solving. Following this, we present a comprehensive case study of Li Auto, a prominent player in the IEV industry. This case study focuses on the company's rapid growth over the past three years, offering insights into its management strategies and the driving forces behind its success. The chapter culminates with a detailed analysis of business decision-making processes for startups in the IEV sector, providing valuable lessons and strategies for navigating this emerging and dynamic market.

5.1. Cultural Clashes in the Making of IEVs

The term IEV naturally signifies the integration of intelligence into the car making process [1]. Traditional car engineering is already a very complex process. The making of IEVs adds additional complexities as it requires talents from many other industries, such as Internet services and software engineering. These talents all come from very different backgrounds with a wide spectrum of work cultures. The mingling of these cultures presents a unique challenge to IEV OEMs, and if not well managed, could lead to failures of product launches, or even worse, safety problems.

In this section, I make an attempt to explain the cultural clashes in the making of IEVs and to provide a few recommendations based on my own experiences. In the past two decades, Internet and software companies, such as Google, Facebook, Amazon, and Microsoft, have dominated the technology world. Technology talents from these Internet companies have spilled over to other industries and made a significant impact. The IEV industry is no exception as it demands talents from these Internet companies to bring innovations into IEVs. Many of these Internet talents are accustomed to fast-paced iterative innovations, such that new product ideas pop up daily and features are defined and evolved quickly. This innovation culture has enabled many Internet startups to expand rapidly through iteratively adapting their products to fit the needs of their target users. For instance, a simple A/B testing would enable product managers and developers to quickly decide whether a feature is beneficial or not.

Nonetheless, in the IEV business, things have changed. Whether it is a startup or an established company, you have only one chance. When you launch the car, either you make it or fail it. You do not have the capital or time to continuously refine a car to meet customer needs. Hence, adapting an Internet-style innovation culture could lead to disasters in the IEV business.

On the other hand, car making is a complex engineering task and its core culture emphasizes stability, reliability, and safety. In the early stage of car definition, everything needs to be defined and not changed afterward, and everyone needs to follow the same schedule. Missing a deadline pretty much means the end of the world in the car business. In addition, to ensure reliability, car OEMs often collaborate deeply with established suppliers, and more reliable components always prevail over more innovative components. To ensure safety, long and comprehensive tests have to be conducted before a car can be released to market. "Oh? You have a new feature idea for the car when I am testing the car? Are you kidding me? I am not going to lose my job because you want to test a new feature." Continuing with this development culture will guarantee that your company gets phased out by newcomers in the IEV business.

As you can expect, when these two groups of people work together, war starts. The software developers would accuse the car engineers of a lack of understanding of innovation and living in the fossil age. Conversely, the car engineers would accuse the software developers of being naïve regarding engineering complexity. Soon, these people no longer talk to each other and your IEV product turns out to be a weird animal.

Some argue for decoupling car engineering and software engineering through exposing a commonly agreed-upon interface. Of course, this would be ideal, but we are entering the stage of scenario-driven car making such that a feature may start from a human–machine interface (HCI) but end in a controller in the chassis. The software developers and the car engineers have to work closely to make this happen.

While I am also eager to develop a formal methodology to address this problem, at this point I can only provide a few practical suggestions. First, I believe the role of project management is crucial to the success of an IEV product. A few years back, I would agree that world-class software developers were key to the success of an IEV product, especially with Tesla's success in integrating many advanced software features into IEVs. After going through the process myself, I believe Tesla had some great project managers, likely Elon Musk himself, who understand both software innovations and car engineering and who are able to develop reasonable goals and judge along the IEV making process what features could be reliably and safely integrated into the product. After all, integrating intelligence into EVs adds complexity in engineering management and hence demands a new generation of project managers who are not there just to enforce a process but also have a deep understanding of both worlds.

Second, let us face it, IEVs have a ten-year life cycle and not all great features can be presented at product launch. Hence, make sure you overprovision your products with enough compute power and sensors so that future features can be smoothly integrated into the IEVs without jeopardizing existing features. In addition, ensure that you have a great over-the-air (OTA) update team to avoid software update disasters.

Third, regarding organizational design, although some teams are decoupled in the traditional car making business, for example the software team and the electrical and electronic (E/E) architecture team, in the making of IEVs, you want to make sure that these teams report to the same boss; otherwise, near product launch your software may not fit in your hardware, and you have to rush to your poor Tier 1 supplier for firefighting.

The cultural clash problem discussed in this section not only applies to IEVs as technology products become more complex. Hopefully, as more IEV products reach the market, a more formal methodology can be developed to manage the ever-increasing complexity of technology products. This methodology by itself would be a great feat of technology and engineering management.

Case Study: Product Strategy and Project Management at Li Auto

Li Auto Inc., based in Beijing, China, has emerged as a prominent contender in the IEV market. Established in 2015, the company has left a lasting mark with its innovative IEVs, specifically the Li ONE and Li L series. In a landscape teeming with next-generation IEV companies in China, such as XPeng and NIO, Li Auto stands out as the first to achieve profitability, a remarkable feat. In this section, we will explore Li Auto's product strategy and project management approach, shedding light on how the company has cultivated a highly efficient organization tailored to the IEV industry.

How Li Auto Turned Profitable

Li Auto became profitable through a combination of strategic product offerings, efficient operations, and wise financial management. Here are the key factors contributing to its profitability [2]:

- **Product Strategy and Sales Growth:** Li Auto secured third place in sales among new energy vehicle (NEV) brands priced over RMB 200,000 in China in the first quarter of 2023. This success is attributed to the strong user endorsement of its vehicle models and the rapid scale-up of its latest offerings. Li Auto's ability to design and produce popular models, combined with an efficient supply chain, manufacturing, and sales network, has been a cornerstone of its financial success.
- **Revenue and Sales Performance:** Li Auto's revenues witnessed a substantial increase of 96.5% year-over-year in the first quarter of 2023, reaching US$2.74 billion. This growth was primarily due to increased vehicle deliveries and the higher average selling price of the Li L series. The company's focus on delivering high-quality vehicles and maintaining strong sales momentum has been crucial in driving its revenue growth.
- **Cost Management and Gross Margin:** Despite the increase in costs associated with higher vehicle deliveries and a varied product mix, Li Auto managed to maintain a healthy gross margin of 20.4% in the first quarter of 2023. This margin reflects the company's ability to manage production costs effectively while scaling up its operations.
- **Operating Expenses and Efficiency:** Li Auto's operating expenses in the first quarter of 2023 were US$498.7 million, a decrease from the last quarter of 2022. This decrease is indicative of the company's operational efficiency. The increase in research and development (R&D) expenses was driven by investment in expanding product portfolios and staff compensation, reflecting the company's commitment to innovation and growth.
- **Profitability and Cash Flow:** Li Auto reported a net income of US$136.0 million in the first quarter of 2023, a significant improvement from a net loss in the same period in 2022. This profitability turnaround can be attributed to strong sales performance and efficient operations. Additionally, the company's strong cash flow, with net cash provided by operating activities at US$1.13 billion in the first quarter of 2023, demonstrates its financial health and ability to generate positive cash inflows from its core business activities.

- **Investments and Nonoperating Income:** Initially, Li Auto's profitability was also aided by gains from interest income and investments, supplementing its core business income. These nonoperating incomes, along with its unique approach to offering extended-range electric cars, helped the company manage costs effectively and cope with supply chain challenges. However, it is important to note that such nonoperating incomes can be variable and subject to changes in the market and policy environment.

Overall, Li Auto's path to profitability has been characterized by strong sales growth, efficient operational management, strategic product development, and effective cost control. These factors, combined with the company's focus on innovation in the EV market, have positioned it well for continued financial success.

Fifteen Challenges Faced by Li Auto

Now we review the challenges during the establishment of Li Auto as stated by Li Auto's founder Li Xiang himself [3].

- **Challenge 1: Market Entry:** There were nearly 400 newly registered automotive manufacturers back in 2018. Even with all the news coverage and information overload, this number was staggering. At that time, BYD had only achieved an annual sales volume of 520,000 units, and foreign automakers still dominated the mainstream and high-end markets. Moving from being one of the 400 newcomers to becoming a leading force among domestic new entrants, even if it's a phase of success, remains an intriguing journey.
- **Challenge 2: Brand Positioning**: Branding holds significant meaning for consumers, especially in the case of automobiles, which are considered substantial assets for ordinary households. Consumers typically aim to achieve two goals through a brand: identity and trust. This transformation is a constant topic of discussion in the automotive industry because a brand's decline can lead to irreversible consequences.
- **Challenge 3: Aligning Diverse Team Members' Ideas:** The automotive industry had enjoyed steady growth for over a decade until around 2015, when the rise of IEV prompted substantial competition as well as salary increases, attracting talent from various fields outside of traditional car making. These newcomers often had different perceptions of their respective brands, which, coupled with steep salary hikes, led to increased egos. Integrating such a diverse group of individuals with varying backgrounds, experiences, and ideologies remains a major challenge, one that continues to evolve.
- **Challenge 4: Establishing Uniform Standards for Good Products:** This question may seem unfamiliar to those who have grown in the automotive industry over the past decade. Standards, processes, and systems appear mature, and technology seems to have reached its peak. For local automakers, leading foreign automakers, and renowned Tier 1 companies, there is a clear guiding target. However, with the advent of the IEV era, how do we address this issue?

- **Challenge 5: Selecting and Developing Teams:** The rise of the IEV industry demands new talents from a wide spectrum of backgrounds, as many fields are new, and there are no ready-made experts. How do we choose talents from traditional automakers and Internet companies? How much automotive knowledge is required, and how much software knowledge is necessary? If we cannot find the right people, can we cross-functionally hire and train them?

- **Challenge 6: Designing Cars for the Market Three to Five Years Ahead:** This question is even more prominent for traditional automakers with a slower pace. It is possible that models planned this year, intended for release in the following years, have already been on the market for two years in the IEV industry. Regardless, staying ahead or at least keeping up with the market is a tremendous challenge.

- **Challenge 7: Identifying Genuine Market Demands:** This challenge is closely related to the previous one. Design should be driven by demand, and demand ultimately depends on the market and users. Can instinctive user research be effective, or will it remain a mere formality?

- **Challenge 8: Managing Over 20,000 Components**: The essence of this challenge lies in making choices. Thousands of components offer many options. How do we make decisions and maintain balance? This has been a persistent theme in the automotive industry, and it is particularly prominent for the IEV industry.

- **Challenge 9: Pricing:** If we were to ask which attribute of a car is the most influential, we could say performance and quality, or we could mention design and specifications. However, when it comes to pricing, everything else takes a back seat. Different groups require different price points. Pricing is the most potent and enduring factor.

- **Challenge 10: Turning Team's Occasional Success into Consistency**: This kind of success can be a blockbuster vehicle or a high-quality delivery from a software team, regardless of size. It requires retrospective analysis, summarizing best practices, internalizing them as know-how, and making them an integral part of the organization.

- **Challenge 11: Planning the Product Matrix:** This challenge differs for traditional car makers and IEV makers. The formers are already struggling with dozens of product lines, including positioning multiple brands and various drivetrain series such as fuel, EV, hybrid EV (HEV), and plug-in hybrid EV (PHEV). Software presents fast iterations and multiple branches. How do we manage and plan this? It is a significant pain point for traditional automakers not adept at software. Most IEV companies are still considering and advancing just a few vehicle models, avoiding the bucket-of-water-on-the-head scenario faced by traditional automakers.

- **Challenge 12: Upgrading Processes for Large Teams:** This challenge varies for new and old forces. IEV companies are like newly recruited construction crews, starting from laying the foundation, while traditional automakers are standing on old-style high-rise buildings. Traditional car makers might now lead in the market, but the IEV industry is developing rapidly; coupled with traditional car makers' shaky foundations, they must transform to survive.

- **Challenge 13: Meeting Customers After Production**: This pertains to how a car reaches consumers' views after leaving the factory, increasing visibility, and improving conversion rates. In this age of intense competition, is the effort put into delivery experiences worth it? This is a question we need to consider.
- **Challenge 14: Generating Profit:** As mentioned earlier, pricing is eternal, but behind pricing always lie costs and profits. The automotive industry, with its high cost sensitivity and thin profit margins, must continuously consider different options to maintain profits.
- **Challenge 15: Ensuring Continuous Growth for the Team and Individuals:** Growth pertains to both individuals and organizations. This challenge remains to be explored in the decades to come.

The Li Auto Management Philosophy

To address the above challenges, Li Auto has developed a set of management philosophies to sustain its growth and profitability in the fierce IEV industry competition.

1. Focus on User Value, Surpass User Needs: This principle emphasizes understanding the underlying needs and desires of users rather than just their expressed requirements. An illustrative example is the transition from corded to mobile phones. Users initially wanted longer cords for their landline phones to move freely while talking, but what they truly desired was the freedom to move without physical constraints, a need that mobile phones fulfilled. This case exemplifies the importance of discerning the fundamental issues or tasks that users seek to address, rather than just providing simplistic features. Product managers often mistake users' proposed solutions for their actual needs. The real need is not the drill but the hole in the wall, or even more profoundly, the picture they want to hang. At Li Auto, the approach is to creatively fulfill these underlying needs by delving into the deeper motivations behind user suggestions.

2. Treat the Organization as a Product: This involves nurturing the organization's growth to support product development. The strategy includes combating laziness, inertia, and ignorance within the team structure. Overcoming laziness is achieved through good work habits at the individual level and efficient processes at the organizational level. Breaking inertia involves disregarding past or conventional methods, focusing instead on what is essential to exceed user expectations. Confronting ignorance is crucial when exploring new possibilities in product development, especially in uncharted territories. The challenges here stem from not knowing what one does not know. To address this, it is important to look beyond the automotive industry for inspiration and to foster a culture of positive reflection within the organization.

3. Positioning—How to Accurately Identify the Product's Target Audience: A key insight presented is the consideration of the life stages of potential customers. The needs people have for a vehicle vary significantly across different stages of life, from starting a career, to establishing a family, raising children, and caring for elderly parents. **By mapping life stages against price points, a new method for segmenting the car market emerges.** However, finding a unique market position is not sufficient. This position must align with a rapidly growing market segment. As a new entrant in the industry, success is unlikely in a shrinking market. Therefore, it is crucial to focus not just on the current size of a market segment but also on its growth trajectory. A rising market segment, even if now smaller, is more valuable than the largest one saturated with competition.

Li Auto's approach involves three key steps:

- Differentiation: New entrants should avoid directly competing for positions already established by others, as catching up can be extremely challenging.
- Market space: Look beyond the current size of market segments and focus on their growth trends. Priority should be given to emerging markets that are experiencing rapid growth, rather than those already crowded with competitors.
- Opportunity identification: It is important to identify areas where customer needs are not fully met by existing products, offering room for innovation. This creates opportunities for new entrants to make a significant impact.

4. Branding—How to Establish Product Consensus with Target Customers?: Shifting from the common misconception that a brand is merely a reputation or a by-product of a good product, a brand is essentially a consensus built with users, centered around three questions: who we are, whom we serve, and what value we create for them. First, defining "who we are" involves considering not just aspirational identities but also recognizing and leveraging competitive strengths. Second, identifying "whom we serve" requires a focused approach, considering four key aspects: the true target audience for the product, the growth potential of this audience, their purchasing power and willingness, and the feasibility of reaching them. This multifaceted analysis helps clarify the served demographic. Third, when addressing "what value is provided," it is critical to distill the answer into a clear, unambiguous statement. Finally, defining a brand is not enough; it must be perceptible in product design and user engagement to truly resonate and build consensus with the target audience.

5. Corporate Culture in Unifying a Diverse Team: The essence of culture in a business context, often perceived as abstract, goes beyond merely enhancing product quality; it fundamentally determines the baseline of the product and the team's performance in crucial moments. True culture is not about filling gaps but is distilled from the team's collective experiences, emphasizing and strengthening the aspects that have previously led to success.

Genuine culture emerges from the behaviors and principles that have historically guided the team toward success. These principles include prioritizing consumer interests, choosing the right course over the easy one, and focusing on achieving a basic level of competency before aiming for perfection. By aligning individual beliefs and actions with these guidelines, a team can significantly boost its collaborative efficiency.

Key takeaways include the following: culture should not be forcibly defined at a team's inception but should evolve and be distilled over one to three years, based on emerging common traits. Culture is about refining the elements that have contributed to past successes, not just compensating for what is lacking. Culture should be cocreated by the team, ensuring it is a shared consensus rather than a directive from the top.

The real test of culture lies in its application to specific situations, with leaders' actions speaking louder than their words. Repeated discussions and debates help the team develop a nuanced understanding and emotional consensus on cultural values. Presenting choices rather than just right-or-wrong questions can clarify issues, as consensus is more readily achieved through choices.

In essence, while branding establishes a consensus with customers externally, culture is about achieving a consensus internally within the team. This balance is vital for a company's success.

6. Quality of a Product: There are three key dimensions in IEV product quality: a sense of safety, a sense of value, and a sense of aspiration. Safety is fundamental in product design, ensuring that users feel secure and protected when using the product. Value pertains to the functionality and experience provided to the target audience for their money. It is not about the manufacturer's perceived importance but about whether the product offers the best possible utility and experience for the user's investment.

Aspiration focuses on the product's ability to evoke desirable scenarios in the minds of users. Achieving a sense of aspiration hinges on a deep and authentic understanding of the user's usage scenarios, aligning the product closely with the essence of what users truly desire.

Breaking industry inertia requires product developers to shift from an industry-centric to a user-centric perspective. **This shift involves immersing oneself in the real usage scenarios of the product**, making it easier to identify and address the unreasonable aspects inherent in industry norms.

7. Creating a Highly Efficient and Collaborative Product Team: Three key actions are required to achieve this: establishing uniform standards, embedding processes, and fostering maturity among team members.

The "Product Four-Step Method" is crucial in this process. This method begins with understanding the user—who they are and what their real needs are. Next, it is important to clarify the company's own goals and aspirations. Then, specific targets and plans aligned with these goals are set. Finally, the outcomes are estimated, and work is planned backward from these expected results.

Choosing the right tools is also essential. Tools must not only be effective but also align with the team's culture and expectations. The tools embody values and should be selected based on the team's ethos to enhance mutual trust and identification within the team. Embedding these standards and processes into daily operations raises the baseline performance of every team member.

Professional skills are important, but maturity is crucial. A team member, regardless of their technical expertise, must be a responsible, self-driven adult capable of defining goals and challenges.

At Li Auto, the emphasis is on recruiting problem-solvers who demonstrate adult maturity. This approach is supported by training systems like the "Seven Habits of Highly Effective People," ensuring that each member contributes positively to the team's collective success.

8. Creating a Product Experience that Exceeds User Expectations: To deliver a superior product experience, it is essential to identify opportunities where competitors have not fully met needs or where existing solutions can be improved. The goal is not merely to stack features or compete on specifications but to judiciously use resources to maximize the enhancement of user experience.

Integrating new technologies can excite users and upgrade their experience. However, if these technologies introduce more problems than they solve, they can ultimately reduce the overall experience. Thus, the successful integration of new technology must consider its net effect on user experience.

The effectiveness of a product is also measured by how frequently users engage with it. The more a product is used, the greater the value it creates for its users. This principle of focusing on high-frequency use is complemented by carefully assessing and targeting areas where competitors fall short, investing more effort in these aspects.

In summary, creating an exceptional user experience requires balancing the pursuit of innovative opportunities with a practical approach. It is about enhancing the user experience in meaningful ways without introducing additional complexities or inconveniences. The ultimate goal is to deliver a product that not only meets but surpasses user expectations through its frequent and effective use.

9. Uncovering the True Needs of Users: The famous quote from Henry Ford about people wanting a faster horse rather than a car illustrates the challenge: users often cannot articulate future needs or innovations. The key is not to rely solely on direct questioning in one-on-one interviews, as these can lead to guided responses and miss unanticipated needs. Instead, the focus should be on cocreation sessions with multiple users, allowing for more open and diverse expression. In these sessions, users should be encouraged to recall and discuss specific scenarios rather than being asked directly about their needs or to propose specific solutions.

This approach shifts the emphasis from rational solutions to emotional expectations. The purpose of cocreation sessions is not to draw immediate conclusions but to enable full and free expression of ideas and experiences. The convergence of ideas and final decision-making are tasks for the product team to undertake afterward, based on the insights gathered.

Li Auto's approach to uncovering real user needs involves two key tactics: first, favoring cocreation over one-on-one interviews; and second, observing more and asking less. After gathering user input, it is crucial to validate the findings. Validation involves assessing the realism of the scenarios and the thoroughness of their implementation.

In summary, effectively identifying user needs requires moving beyond traditional user research methods. By engaging users in cocreation and focusing on observation, a more authentic and diverse range of needs can be uncovered. This approach allows for a deeper understanding of what users truly want and need, leading to more successful and user-centric product development.

10. Selecting the Appropriate Technological Path: Technology should serve the product's value. The first principle in choosing technology is to ensure it enhances the product and user value. This means making choices that are specific to the targeted user group, rather than a general audience, and avoiding the temptation to chase novelty or trends just for the sake of being current.

The second principle involves leveraging the strengths of different technologies, combining them to complement each other. This approach is not about choosing one technology over another in absolute terms but about finding a synergistic blend. It is important to learn from both the successes and failures of others, analyzing the underlying conditions and key issues behind these outcomes to find innovative opportunities.

Lastly, the choice of technology should also align with the company's resources and capabilities, considering cost and efficiency. However, it is not about minimizing costs but optimizing efficiency—the goal is to create a product that offers users greater value at a lower cost.

11. Setting Competitive Prices: There are three key aspects in achieving this: rational product value, emotional user perception, and the company's operational needs. Rationally, product value is a function of product strength, price, and quality. Stronger product features, lower prices, and higher quality all contribute to better sales.

However, setting a price is not just about these objective factors; it also involves understanding the user's emotional response. From an emotional perspective, the psychological price point is crucial. It does not solely depend on the cost of the product but also on how users perceive it: as a functional item, an emotional attachment, or a symbol of status. The same type of product can command different prices based on its positioning in the user's mind. For products like cars, which often symbolize status, pricing must align with users' psychological price points.

Additionally, the company's operational needs must be considered. A healthy gross margin and the ability to sustain operations are vital. For instance, maintaining a 20% gross margin is an essential factor in pricing decisions for a sustainable business model.

Pricing should reflect the strongest product value within the price range, align with the nuanced feelings of users, and ensure the company's sustainability and profitability. This approach ensures that pricing decisions support both market competitiveness and long-term business viability.

12. Review and Improve Iteratively: Transforming successful products from occasional hits into consistent outcomes requires effective postproject reviews, which depend on controlling three aspects: the objectives, the process, and the organization of the review sessions. The primary goal of these reviews is to convert experiences into capabilities, focusing on future improvements rather than past performance. It is not about assigning blame or credit but about reaching a consensus on how to do better in the future.

For a productive review, it is essential to compare the outcomes against the original plan. This involves a thorough PEA (plan, execute, assess) analysis at the planning stage, clarifying six elements:

- Problem definition: The initiator must provide concrete evidence that they are addressing a real problem. Proposals failing to demonstrate this are rejected.
- Objective requirements: Clearly define the target user without ambiguity and specify the desired user experience.
- Solution description: Include both the functionality postimplementation and the detailed plan to achieve it.
- Risk assessment: Identify potential risks and have contingency plans in place.
- Resource evaluation: Assess R&D resources, costs, licensing fees, and development timelines.
- Operational metrics: Establish quantifiable operational metrics with a direct relationship to the set objectives and predetermined expectations for each metric.

To summarize, the first priority in postproject reviews is to identify and consolidate the strengths of the previous product, ensuring that the review session is constructive rather than critical. Reviews should be aligned with planned objectives and metrics, using tools like Li Auto's PEA to set clear targets and indicators beforehand. This approach gives reviews a focused direction. Lastly, the success of these sessions heavily depends on the leadership, particularly the CEO or the primary leader, who must strictly adhere to review protocols to set an example and guide participants effectively. This disciplined approach ensures that each review session is purposeful and contributes to continuous improvement and consistency in product success.

13. Strategies for Scaling a Product: In the startup phase (from 0 to 1), the focus should be on product validation and establishing a minimal viable business cycle. At this stage, without the advantages of industrial scale or technological leadership, the key to moving forward is to maximize the strengths of the product to the highest industry standards. This approach is seen as the quickest path to securing a position in the next phase of growth.

As the product moves from 1 to 10, the approach shifts. It is no longer sufficient to follow market inertia; instead, it is crucial to anticipate the future market size and identify the turning point—a critical moment when winners take all, and losers exit the market. Setting goals in this scale-up phase requires a shift from inertia to necessity. The new logic in goal-setting involves preparing for future market challenges and opportunities.

The principle is to overprepare rather than underprepare. While overpreparation might lead to some resource waste, this waste is limited compared to the risks of being underprepared. In hindsight, conservative strategies and resource allocations often seem insufficient.

In summary, the journey from 0 to 1 requires focusing on turning strengths into industry benchmarks and concentrating resources on significant initiatives. Moving from 1 to 10 demands a change in goal-setting logic, from inertia-driven to necessity-driven, ensuring survival in a competitive market. In terms of resource allocation, it is better to overallocate than to fall short, as the cost of underpreparation in a highly competitive environment can be far greater than the limited waste resulting from overpreparation.

<u>14. Establishing a Process for Systematic Product Development</u>: This strategy emphasizes different strategies for the initial growth phase (from 0 to 1) and the scale-up phase (from 1 to 10). In the early stage of a product or company, when there are fewer products and a smaller team, a rapid and flexible approach is more effective. However, as the product range expands and the company grows into a large-scale operation, a more structured and process-driven approach becomes essential. At this stage, inefficient processes can significantly hinder the ability to continuously create successful products.

The concept of a "user value stream" is central to refining processes. It involves repeatedly asking who the users are and understanding their experiences with the product. This understanding should then dictate all internal processes, ensuring they align with delivering user value. By focusing on the user value stream, necessary processes can be identified and streamlined, enhancing overall efficiency. Best practices should be standardized and embedded into these processes.

Li Auto's recent experience in process development highlights three main benefits:

- Building a highway for complex business: Though highways have the most rules, they also offer the highest efficiency and lowest accident rates. This analogy suggests that well-designed processes can significantly improve the efficiency of creating and delivering value to users.
- Establishing a robust training system: With basic training, every team member, including new recruits, can quickly learn to navigate these processes. This system allows for rapid growth and skill acquisition, eliminating the need for traditional apprenticeship models.
- Creating an interlocking mechanism: This approach represents a fundamental shift from alignment (as with Objectives and Key Results, or OKRs) to interlocking, transitioning from a trust-based system to a contract-based one. It fosters mutual challenge-setting and close collaboration among teams, aligning individual and team goals for optimal cooperation and shared success.

In summary, as a product or company grows, it is crucial to evolve the approach to process management. For Li Auto, this evolution has led to the creation of efficient, interconnected processes that streamline business operations, enhance training and growth, and foster better team collaboration, all aligned with delivering user value.

<u>15. Managing Physical Stores</u>: This strategy addresses why physical stores remain crucial for selling low-frequency, high-value products like cars. For customers, the decision-making process for these types of purchases is often long and complex, necessitating multiple experiences with the product before making a purchase. This is particularly true for new brands. The essence of sales work is broken down into two main modules: customer acquisition and conversion.

Customer acquisition involves strategies like opening new stores, promotions, events, outreach, and buying sales leads. Conversion tactics include price reductions, sales promotions, hiring, and enhancing incentives. To enhance the efficiency of direct sales stores, there are two core methods:

- Rapid store expansion: This involves reducing the variety of stock-keeping units (SKUs), establishing a training and examination system, and preparing a robust digital system.
- Balancing control and autonomy: While maintaining control over direct sales stores, it is also crucial to grant autonomy to individual stores. This involves cultivating a sense of ownership and entrepreneurial spirit among the regional managers and store managers.

In summary, improving sales efficiency in direct stores requires a combination of strategic expansion and operational autonomy. Fast store openings with streamlined product offerings and well-prepared digital systems can facilitate customer acquisition. Meanwhile, empowering store managers to make decisions and manage their stores promotes effective conversion. This balanced approach enables brands, especially new ones, to more effectively reach and convert potential customers in the high-stakes market of automobile sales.

16. Enhancing Operational Efficiency: This strategy emphasizes the importance of breaking down data into their smallest components. This approach ensures that each element of the business operates healthily, contributing to the overall health of the organization. In the context of Li Auto, an example is given of how to assess the efficiency and cost-effectiveness of a product expert. This involves calculating how many cars they need to sell per month to justify their salary, which is above the industry standard. To determine this, various factors are considered, such as the number of test drives needed based on conversion rates and the amount of customer leads required. This assessment extends to the time needed for managing these leads and arranging test drives, verifying if these align with normal standards. Li Auto emphasizes surpassing user needs, but it always circles back to creating real product value for the user. If resources are used excessively without adding clear value to the product, it is considered a waste or an indication of biased supplier relationships.

Controlling profit margins involves two main points:

- Detailed budgeting and forecasting: Every aspect of the business, including each step in every process and every individual involved, should be atomically split in the budget and forecast. This granularity allows for prompt identification and adjustment of issues during execution.
- Implementing best practices in production: By identifying and sharing best practices, every participant in the production and sales chains can align their activities with these standards. This approach ensures that all aspects of production and sales are optimized and consistently improved.

In essence, improving operational efficiency and profitability requires a detailed, atomic-level analysis of every aspect of the business and the adoption of best practices across all stages of production and sales. This thorough approach enables timely problem-solving and consistent optimization of processes, leading to enhanced efficiency and profitability.

5.2. Critical Business Decision-Making for IEV Startups

Most business decisions are made with analysis, but some are judgment calls not susceptible to analysis due to time or information constraints. In this section, we present a real-life case study of critical business decision-making of PerceptIn, an autonomous driving technology startup: in the early years, PerceptIn had to decide on the design of computing systems for its autonomous vehicle products [4]. By providing details on PerceptIn's decision-making process and the results of the decision, we hope to provide some insights that can be beneficial to entrepreneurs and engineering managers in technology startups.

Background: PerceptIn was established in 2016 to develop visual perception technologies for autonomous vehicles and robots. Since its inception, PerceptIn has successfully attracted more than 10 million USD of venture capital funding, from Walden International, Matrix Partners, and Samsung Ventures. PerceptIn is an international technology startup with operations in the United States, Japan, Europe, and Asia. PerceptIn consists of more than 30 researchers and engineers and ten business professionals. The business professionals are responsible for business development in different markets and gather feedback for the company's R&D efforts, whereas the engineers and researchers are responsible for developing cutting-edge autonomous driving technologies. In the past three years, PerceptIn has generated over 20 US patents and over 100 international patents, as well as many research papers.

In 2017, PerceptIn decided to develop low-speed autonomous vehicles to serve the micromobility market, as micromobility is a rising transport mode wherein lightweight vehicles cover short trips that massive transit ignores [5]. According to the US Department of Transportation, 60% of vehicle traffic is attributed to trips fewer than 5 miles. Transportation needs for short trips are disproportionally underserved by current mass transit systems due to high cost, which affects the society profoundly. Micromobility bridges transit services and communities' needs, driving the rise of mobility-as-a-service.

Business Analysis: PerceptIn's primary customers are autonomous vehicle operators around the globe, and PerceptIn partners with these autonomous vehicle operators to provide micromobility services in different markets, such as Japan, the United States, and Europe. PerceptIn's ultimate goal is to provide affordable and reliable autonomous driving technologies that can allow PerceptIn's operators to generate profits and subsequently grow the business. In 2017 and 2018, PerceptIn conducted over ten pilot projects globally to understand the micromobility market, the customer needs, as well as the cost structure of this business.

One of PerceptIn's pilot projects took place in 2017 at ZTE's industrial park in Shenzhen, China. ZTE is a leading Chinese telecom company with an enormous campus in Shenzhen, and the campus is filled with over 30,000 workers with tremendous intracampus transportation needs. In this pilot project, PerceptIn's pods transferred ZTE's workers across the campus. Each PerceptIn's pod packs four HD cameras, four mid-range radar sets, and ten ultrasound sensors, as well as GPS and sensors for wheel odometry. From these pilot projects, PerceptIn collected sufficient operation data and customer feedback for internal business analysis.

Based on internal business analysis, if PerceptIn can provide low-speed autonomous vehicles under $70,000 per unit, PerceptIn could generate a reasonable return on investment (ROI) for PerceptIn's customers, the autonomous vehicle operators. Thus, in 2018 PerceptIn set the goal to develop autonomous vehicles that can be sold at a price tag of $70,000, which is five to ten times lower than what is commonly believed to be possible for commercial autonomous vehicles. However, the $70,000 price tag also imposes very strict and challenging constraints on the design of low-speed autonomous vehicles. In detail, we have to break down the $70,000 into nonrecurring engineering (NRE) costs such as R&D, recurring costs such as the cost of the chassis, the cost of drive-by-wire conversion (meaning to convert a traditional vehicle into one that can be controlled by computers), the cost of sensors, the cost of integration, the cost of customer service, and finally the cost of the computing system [6].

Situation: In June 2017, based on the initial feedback from the ZTE case study, PerceptIn conducted a study on autonomous driving computing systems [7]. PerceptIn concluded that computing is the bottleneck for the commercial deployment of autonomous vehicles, and PerceptIn needed a computing system that is reliable, affordable, high-performance, and energy-efficient. Most importantly, we needed a solution that is cost-effective and has a short time-to-market. PerceptIn faced several options:

1. Optimization of commercial off-the-shelf mobile system-on-chip (SoC) computing systems: This approach brings several benefits: first, since mobile SoCs have reached economies of scale, it would have been most beneficial for PerceptIn to build its technology stack on affordable, backward-compatible computing systems. Second, PerceptIn's vehicles target micromobility with limited speed, similar to mobile robots, for which mobile SoCs have been demonstrated before. However, an extensive study is required to fully understand mobile SoCs' suitability for autonomous driving. This may delay PerceptIn's product launch by six months.
2. Procurement of specialized autonomous driving computing systems: There were commercial computing platforms specialized for autonomous driving, such as those from NXP, MobilEye, and NVIDIA. They are mostly application-specific integrated circuit (ASIC)-based chips that provide high performance at a much higher cost. For instance, the first generation of Nvidia PX2 system costs over $10,000. Besides the cost issue, these computing systems mostly accelerate only the perception function in autonomous driving, whereas PerceptIn requires a system that optimizes the end-to-end performance.
3. Development of proprietary autonomous driving computing systems: Developing a proprietary computing system guarantees that PerceptIn has the most suitable system for PerceptIn's customers and for its workloads, but also means that PerceptIn needs to invest a significant amount of financial and personnel resources on this project. Also, the investment does not guarantee the success of this project. It is a huge and risky bet for a startup like PerceptIn.

Decision: Then the decision-making process started. Without quantitative evaluation methods, we had to use judgment calls to evaluate different options [8, 9]. We summarize these options in Table 5.1. The parameters of evaluation include affordability, backward compatibility, suitability for autonomous driving computing, and project risk. For all parameters, the higher the better. For affordability, option 1 is the clear winner. For backward compatibility, both option 1 and option 3 deliver good results. For suitability, option 3 is the clear winner, and option 1 is unknown and requires additional study. For risk, both option 1 and option 2 have low risk.

Using Table 5.1, PerceptIn quickly ruled out option 2 due to its low score across all parameters except suitability. It is clear from available data that option 2 was the worst choice among the three. Then, an internal debate began within PerceptIn on whether to move forward with option 1 or option 3. Option 1 seemed very attractive, but it was unknown whether it would be suitable for autonomous driving computing tasks. If it was not suitable, then option 1 would be infeasible. To have a clear answer on whether option 1 would be suitable, a six-month study would be necessary. For option 3, it would be an extremely expensive option and with high risk. Even if PerceptIn invested in this project, there would be no guaranteed success.

Table 5.1 Comparisons of the three available options.

	Affordability	Backward compatibility	Suitability	Risk
Option 1: mobile SoCs	HIGH	HIGH	UNKNOWN	LOW
Option 2: specialized computing systems	LOW	LOW	MEDIUM	LOW
Option 3: proprietary computing systems	LOW	HIGH	HIGH	HIGH

Since there is an unknown parameter in option 1, a deterministic analysis could not be conducted, and the PerceptIn management team had to use judgment calls not susceptible to analysis in our decision-making process.

After several rounds of internal debate, the PerceptIn management team decided to compare option 1 and option 3 in their respective worst-case scenario. If PerceptIn moved forward with option 1, the worst case that could happen was that six months were wasted, but with limited investment. If PerceptIn did find out that option 1 would not be suitable, PerceptIn could still try option 3. If PerceptIn moved forward with option 3, the worst case that could happen was that half of PerceptIn's R&D budget and 12 months were wasted. With this analysis, PerceptIn decided to take the safest approach, also an approach that everyone was comfortable with: move forward with option 1 for six months; if that does not work, then try option 3.

Results: For six months, PerceptIn focused on option 1, but unfortunately PerceptIn found that mobile SoCs are ill-suited for autonomous driving for three reasons:

1. The compute capability of mobile SoCs is too low for realistic end-to-end autonomous driving workloads. Figure 5.1 shows the latencies and energy consumptions of three perception tasks— depth estimation, object detection, and localization—on an Intel Coffee Lake CPU, NVIDIA GTX 1060 GPU, and NVIDIA TX2, which represents today's

high-end mobile SoCs. Figure 5.1a shows that TX2 is much slower than the graphics processing unit (GPU), leading to a cumulative latency of 844.2 ms for perception alone. Figure 5.1b shows that TX2 has only marginal, sometimes even worse, energy reduction compared to the GPU due to the long latency.

2. Mobile SoCs do not optimize data communication between different computing units, but require redundant data copying coordinated by the power-hungry central processing unit (CPU). For instance, when using a digital signal processor (DSP) to accelerate image processing, the CPU has to explicitly copy images from the sensor interface to DSP through the entire memory hierarchy. PerceptIn's measurement shows that this leads to an extra 1 W power overhead and up to 3 ms performance overhead.

3. Traditional mobile SoCs design emphasizes compute optimizations, while PerceptIn finds that for autonomous vehicle workloads, sensor processing support in hardware is equally important. For instance, autonomous vehicles require very precise and clean sensor synchronization, which mobile SoCs do not provide.

Figure 5.1 Performance and energy consumption comparisons.

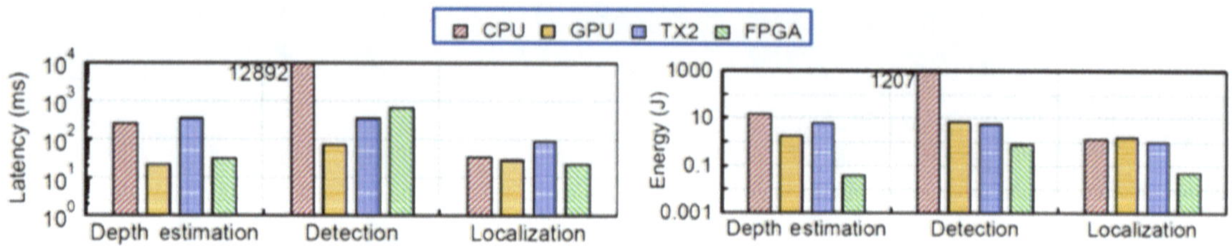

Starting in early 2018, PerceptIn decided to move forward with option 3 since option 1 proved not to work. PerceptIn thus formed a team to develop the field-programmable gate array (FPGA)-based DragonFly computing system [10, 11, 12, 13]. Option 3 was a huge success. Today all autonomous vehicles shipped by PerceptIn are empowered by PerceptIn's proprietary DragonFly computing system. However, if PerceptIn had boldly moved forward with option 3 at the beginning, PerceptIn would have been able to ship this great product six months earlier. The main reason of option 3's success was that when using judgment call, we overestimated the technical risk. Of course, during the development process of option 3, we encountered a lot of technical unknowns, and fortunately, we found solutions to these technical problems. However, during the planning stage, using judgment call, we overestimated the difficulties of these technical problems, which led us to believe that starting with option 1 was a safer approach.

Retrospective: PerceptIn shipped its products globally but delayed by six months because

PerceptIn took an R&D detour. At the planning stage, when using judgment call, PerceptIn overestimated the technical difficulties in option 3 and chose to start with a safe path, option 1. In a way, PerceptIn's evaluation was too pessimistic; had PerceptIn taken the hard choice to go with option 3 initially, PerceptIn would have a much higher market share today.

By making the decision to go with option 3, PerceptIn would have greatly widened its moat and enlarged its edge over competition. In retrospect, the root cause was that the PerceptIn team did not have proper anchoring and adjustment when using judgment call [9], and hence, the decision was too pessimistic.

The lesson from this project is that, for a technology startup in hypergrowth mode, when its management team identifies a critical technical problem, the startup should always choose the best solution over the safest solution. If the team is not comfortable with its decision, outside expert opinion should be enlisted to help provide proper anchoring. This will definitely give the startup a huge advantage over its competitors who might choose the easiest solution.

5.3. Summary

In conclusion, IEV is a burgeoning industry where project management methodologies are continuously evolving. This chapter has highlighted the significant challenge of cultural clashes inherent in IEV production, a process that demands a synthesis of varied expertise and perspectives. Furthermore, we have presented a thorough case study examining the scaling strategies of an IEV business, offering valuable insights into the growth dynamics within this nascent industry. Additionally, we have explored a detailed case study focused on decision-making processes within a technology startup in the IEV domain, shedding light on the intricate considerations and strategic approaches vital for success in this innovative and rapidly changing field.

References

1. Liu, S., "Cultural Clashes in the Making of Intelligent Electric Vehicles," Forbes, accessed June 1, 2024, https://www.forbes.com/sites/forbestechcouncil/2023/03/28/cultural-clashes-in-the-making-of-intelligent-electric-vehicles/.
2. Li Auto Inc, "Announces Unaudited First Quarter 2023 Financial Results," Li Auto, accessed June 1, 2024, https://ir.lixiang.com/news-releases/news-release-details/li-auto-inc-announces-unaudited-first-quarter-2023-financial.
3. Xiang, L., "16 Lectures on Product Practicalities (in Chinese)," accessed June 1, 2024, https://www.bilibili.com/video/BV1kh4y1U7cH/.
4. Liu, S., "Critical Business Decision Making for Technology Startups: A PerceptIn Case Study," *IEEE Engineering Management Review* 48, no. 4 (2020): 32-36.
5. Liu, S. and Gaudiot, J.L., "Autonomous Vehicles Lite Self-Driving Technologies Should Start Small, Go Slow," *IEEE Spectrum* 57, no. 3 (2020): 36-49.
6. Liu, S., *Engineering Autonomous Vehicles and Robots: The DragonFly Modular-Based Approach* (Hoboken: John Wiley & Sons, 2020).
7. Liu, S., Tang, J., Zhang, Z., and Gaudiot, J.L., "Computer Architectures for Autonomous Driving," *Computer* 50, no. 8 (2017): 18-25.
8. Kahneman, D. and Klein, G., "Conditions for Intuitive Expertise: A Failure to Disagree," *American Psychologist* 64, no. 6 (2009): 515.
9. Tversky, A. and Kahneman, D., "Judgment under Uncertainty: Heuristics and Biases," *Science* 185, no. 4157 (1974): 1124-1131.

10. Fang, W., Zhang, Y., Yu, B. and Liu, S., "DragonFly+: FPGA-Based Quad-Camera Visual SLAM System for Autonomous Vehicles," in *Proceeding IEEE HotChips*, Cupertino, CA, 2018, 1.

11. Liu, Q., Qin, S., Yu, B., Tang, J. et al., "π-BA: Bundle Adjustment Hardware Accelerator Based on Distribution of 3D-Point Observations," *IEEE Transactions on Computers* 69, no. 7 (2020): 1083-1095.

12. Tang, J., Yu, B., Liu, S., Zhang, Z. et al., "π-SoC: Heterogeneous SoC Architecture for Visual Inertial SLAM Applications," in *2018 IEEE/RSJ International Conference on Intelligent Robots and Systems (IROS)*, Madrid, Spain, 2018, 8302-8307, IEEE.

13. Yu, B. Hu, W., Xu, L., Tang, J. et al., "Building the Computing System for Autonomous Micromobility Vehicles: Design Constraints and Architectural Optimizations," in *2020 53rd Annual IEEE/ACM International Symposium on Microarchitecture (MICRO)*, Athens, Greece, 2020, IEEE.

Chapter 06

Enabling Technologies—Electrification

Electrification is not just a shift in power sources; it's a renaissance of the automotive industry, sparking innovation in design, efficiency, and the relationship we have with our vehicles.

—Anonymous

Electrification technologies encompass a wide range of innovations designed to enhance vehicle efficiency, performance, and sustainability. These technologies are fundamental to the operation of IEVs and include advancements in battery systems, electric motors, power electronics, and, most importantly, electrical and electronic (E/E) architectures. Electrification technologies are essential to IEVs because they directly address the environmental impact of traditional fossil fuel-driven vehicles by reducing carbon emissions and pollution. Additionally, they offer an enhanced driving experience with lower operational costs due to fewer moving parts and reduced maintenance requirements compared to ICEs. The continual advancement of these technologies is vital for improving the viability, performance, and acceptance of EVs in the global market.

6.1. Battery Technology and Management

IEVs rely heavily on advanced battery technologies that offer higher energy densities and longer lifespans. Research focuses on lithium-ion (Li-ion) batteries due to their favorable energy-to-weight ratios and efficiency. Battery technology and management systems are pivotal in the operation and efficiency of IEVs, serving as the cornerstone for their energy source and usage optimization [1, 2, 3].

6.1.1. Batteries

IEVs predominantly utilize Li-ion batteries due to their high energy density, which allows for greater range and efficiency [4]. These batteries benefit from advancements in materials science, particularly through the development of cathode and anode materials which enhance the battery's capacity and longevity. Innovations such as solid-state batteries are also emerging, offering higher energy densities, improved safety, and faster charging times compared to conventional liquid electrolyte solutions.

A Li-ion battery in an EV consists of many electrochemical cells, each containing a cathode, an anode, and an electrolyte. The cathode is typically made from metal oxides that include nickel, manganese, and cobalt, while the anode is usually made from graphite. The electrolyte is a liquid that facilitates the movement of ions. During the charging phase, electricity from an external source causes lithium ions to move from the cathode to the anode through the electrolyte, storing energy in the process. When the battery discharges, these ions travel back to the cathode, releasing stored energy as electricity that powers the EV's motor. The charge and discharge cycles begin with the anode releasing lithium ions to the cathode while the device is being used, creating a flow of electrons that provides power. When the battery is charged, the flow is reversed: lithium ions move from the cathode back to the anode, ready for the next discharge cycle.

Li-ion batteries are favored in EVs due to their efficiency and high power-to-weight ratio, which allows for a longer range per charge and a relatively fast recharge time. However, the extraction and processing of materials such as lithium and cobalt raise environmental and ethical concerns due to the impacts of mining and the geopolitical issues associated with mineral procurement. Research continues into enhancing the sustainability of these batteries, including the development of technologies that reduce reliance on rare materials and improve recyclability, as well as innovations like solid-state batteries that offer potential improvements in stability and charging speed.

The landscape of Li-ion battery suppliers for IEVs is dominated by a few key players who have established significant competitive advantages through technological innovations and market strategies. CATL holds a leading position in the global market through aggressive capacity expansion, strategic partnerships with major automakers, and advancements in battery technology, such as the introduction of its Qilin and Shenxing batteries. These batteries are known for their high energy density and safety features, which have allowed CATL to maintain and even expand its market share despite intense competition.

BYD is currently at the second position in the market. BYD has benefited from its dual role as both a battery manufacturer and an automaker, which enables better integration of battery technologies into its vehicles. The company has also been expanding its third-party supply arrangements, which include partnerships with major global automakers, thereby broadening its market reach.

LG Energy Solution, which ranks among the top suppliers, has faced challenges due to competition and shifts in client bases, such as General Motors (GM). Despite these challenges, LG continues to be a major player due to its strong technological base and broad portfolio of batteries suitable for a wide range of automotive applications.

6.1.2. Battery Management Systems

Battery management systems (BMSs) are crucial for safeguarding the battery's health and optimizing its performance. They continuously monitor the battery's state of charge (SOC), state of health (SOH), and temperature, ensuring operations within safe parameters [5]. The BMS plays a critical role in extending the battery's life by preventing overcharging and deep discharging, which can degrade the battery cells. Advanced BMSs integrate algorithms for more accurate SOC and SOH estimations, enhancing the reliability of the battery and the vehicle's range predictions.

Key functions of a BMS include the following:

- **Monitoring:** BMS keeps track of key parameters such as voltage, current, and temperature of each cell within the battery pack. These data help in assessing the battery's overall health and performance.
- **Protection:** The system provides essential safety mechanisms, including overvoltage, undervoltage, overcurrent, and overtemperature protections. These functions prevent conditions that could lead to battery damage or safety hazards like thermal runaway.
- **Cell balancing:** Cell balancing is critical for maintaining uniform charge levels across all cells in the battery pack. This not only enhances battery performance but also extends its lifespan by preventing the overcharging or deep discharging of individual cells.
- **State estimation:** BMS calculates key state indicators such as SOC and SOH. SOC indicates the remaining charge in the battery, while SOH reflects the overall condition and efficiency relative to a new battery.
- **Thermal management:** Maintaining an optimal temperature range is vital for battery efficiency and longevity. The BMS manages cooling and heating processes to keep the battery within a specified temperature range, improving performance and safety.
- **Predictive analytics:** BMS forecasts future battery conditions and performance based on historical data and usage patterns. This capability can significantly enhance the operational efficiency and reliability of EVs by preempting potential issues and optimizing charging strategies.

The leading suppliers of BMS technologies are reviewed as follows: Denso Corporation has developed an innovative battery-monitoring integrated circuit (IC) for Li-ion batteries. This IC is notable for its ability to detect battery voltage across multiple cells, significantly enhancing the efficiency of EVs and hybrid vehicles by optimizing battery usage and extending vehicle range. Bosch has been proactive in integrating cloud services for BMS, which allows real-time monitoring and optimization of the battery status. This service enhances user interaction by providing instant data on battery health, which can be crucial for maintaining the longevity and efficiency of EV batteries. Tesla also continues to innovate with its multichannel and bi-directional BMS, which includes features like dynamic redundancy to ensure reliability and safety even when primary systems fail. This technology underscores Tesla's approach to integrating robust safety mechanisms within its BMS. Moreover, companies like Visteon are advancing wireless BMS technologies, which GM is set to incorporate into its EVs. This technology eliminates the need for traditional wired connections, reducing complexity and enhancing the scalability of battery systems.

6.1.3. Thermal Management

Thermal management is vital for maintaining battery efficiency and safety. IEVs incorporate sophisticated cooling systems to manage the heat generated during high-current discharges and recharges. These systems range from air-cooled to liquid-cooled systems, with the latter providing more effective heat dissipation. Proper thermal management not only extends the battery's lifespan but also maintains performance across various ambient temperatures.

Some of the leading suppliers in this field have developed innovative solutions that offer significant competitive advantages. Marelli has introduced an integrated Thermal Management Module (iTMM) that combines the thermal management of the electric drive, battery, and vehicle interior into one compact system. This system is designed to be more efficient and requires less installation space, which could potentially increase vehicle range by up to 20% under winter conditions. Marelli's system uses a smart valve arrangement to connect various heat exchangers efficiently, thereby optimizing energy management and reducing system complexity.

Schaeffler offers a comprehensive range of thermal management components that can be integrated into existing vehicle systems. Their approach focuses on efficient interaction among components to optimize the cooling and heating needs of EVs. Schaeffler's systems are designed to manage the thermal conditions of batteries, electric motors, and power electronics effectively, ensuring that each component operates within its optimal temperature range. This integrated approach is crucial for prolonging battery life, enhancing vehicle performance, and ensuring fast charging capabilities. Bosch provides another example of excellence in thermal management for hybrid and EVs. Their systems are designed to ensure efficient battery cooling, even at high temperatures, which helps in fast charging and extends the lifetime of the battery.

6.2. Power Electronics and Propulsion

The role of power electronics in IEVs is to efficiently convert and control the flow of electrical energy within the vehicle's powertrain. This includes the development of traction inverters and converters, which manage the delivery of power between the battery and the motors. Advances in semiconductor materials, such as silicon carbide (SiC) and gallium nitride (GaN), are pivotal as they allow for more efficient operation at higher temperatures and voltages, which is beneficial for reducing losses and improving the vehicle's range and performance.

6.2.1. Power Electronics in IEVs

Power electronics in IEVs primarily manage the flow of electrical energy within the vehicle. This includes the conversion and control of electrical power from the battery to the drive motors and other vehicle systems [6, 7, 8]. Major power electronics components include inverters and converters. **Inverters** convert direct current (DC) from the battery into alternating current (AC) to drive the electric motors. They are crucial for adjusting the speed and torque of the motor by varying the frequency and amplitude of the AC power. **Converters** include DC–DC converters that step up or step down voltage levels to suit different components within the vehicle. For example, they might reduce the high voltage from the main drive battery to lower voltages required for vehicle accessories and auxiliary systems.

BorgWarner has made significant contributions with its introduction of SiC-based inverters, which are known for their high efficiency and power density. They offer 800 V inverter technology that is scalable and efficient, designed specifically for high-voltage battery EV (BEV) platforms. This technology supports more robust electrical systems and faster charging capabilities, addressing key needs in the EV market.

Onsemi also has entered into a strategic partnership with Volkswagen (VW) to supply components for complete traction inverter solutions on next-generation VW electric platforms. Their EliteSiC 1200 V SiC power modules are pivotal in enhancing the efficiency and performance of front and rear traction inverters, showcasing their commitment to advancing inverter technology in the automotive sector.

Infineon supplies a range of products from insulated-gate bipolar transistors (IGBTs) to newer SiC metal-oxide-semiconductor field-effect transistors (MOSFETs), focusing on high efficiency and reliability. Infineon's products are designed to handle high voltages and currents with reduced losses, which is crucial for achieving longer battery life and better performance in EVs.

Hitachi supplies high-performance inverters that utilize advanced technologies to improve the efficiency of EVs. Their inverters are equipped with features that ensure high-speed operation and durability, making them suitable for a wide range of automotive applications including high-performance and heavy-duty vehicles.

Mitsubishi Electric is known for its high-quality electrical components, including inverters that are widely used in the automotive industry. Their inverters are designed to be highly efficient and reliable, with capabilities that support fast charging and high-power applications essential for next-generation EVs.

6.2.2. Propulsion Technologies in IEVs

Propulsion technologies primarily revolve around electric motors and advanced drivetrain systems. These technologies are central to the functioning of IEVs and distinguish them from traditional ICE vehicles. The main components involved are as follows:

Electric Motors: The most common types used in IEVs are permanent magnet synchronous motors (PMSMs) and induction motors. PMSMs are favored for their efficiency and power density, which are beneficial for automotive applications requiring high performance and compact size. Induction motors, while generally less efficient, benefit from simpler construction and lower costs due to the absence of rare-earth magnetic materials.

Specifically, PMSMs are highly efficient due to their use of permanent magnets, which generate a constant magnetic field without the need for electrical power. This attribute leads to higher efficiency and power density, allowing PMSMs to deliver more power relative to their size. They also provide better control over speed and torque, which is crucial for EVs requiring precise motor control for applications like traction. However, PMSMs are more expensive because of the cost of rare-earth permanent magnets, and they can be sensitive to high temperatures, which might lead to demagnetization.

On the other hand, induction motors are prized for their simplicity and durability. They lack permanent magnets, relying instead on electromagnetic induction to operate. This makes them less sensitive to temperature fluctuations and generally cheaper and easier to produce and maintain. Induction motors are excellent for regenerative braking systems due to their ability to act as generators more efficiently during vehicle coasting or braking. However, they generally have lower efficiency and power density compared to PMSMs and can be more complex to control, requiring sophisticated electronic management systems to optimize their performance under varying loads.

The selection between a PMSM and an induction motor often depends on specific vehicle performance requirements, cost constraints, and intended market positioning. High-performance EVs, where efficiency and precise control are paramount, often utilize PMSMs despite their higher cost. Conversely, induction motors are favored in more cost-sensitive applications where robustness and cost reduction are key, such as in mass-market EVs. Thus, manufacturers weigh these factors when deciding which motor technology to implement in their vehicles, considering both the technical benefits and the economic implications of each motor type.

Leading suppliers of electric motors for IEVs include Bosch, ABB, Nidec, Brose, and Johnson Electric.

Motor Controllers: These are integral to managing the performance of the electric motor. They adjust the amount of power delivered to the motor, controlling speed and torque based on driver inputs and vehicle conditions. Motor controllers play a critical role in converting the battery's DC power into AC power that can be used by the motor, while also regulating the speed, torque, and overall efficiency of the motor based on the driver's input and vehicle conditions.

The core functionality of a motor controller in an IEV revolves around its ability to precisely control the power supplied to the electric motor. This is achieved through sophisticated electronic techniques, including pulse-width modulation (PWM), which adjusts the motor's speed by varying the duration of the electrical pulses sent to the motor. This method helps optimize energy consumption and control the amount of torque being generated, enhancing the vehicle's responsiveness.

Another important aspect of motor controllers is their integration with the vehicle's regenerative braking system. This system allows the vehicle to recover kinetic energy during braking and convert it back into electrical energy, which is then stored in the battery. The motor controller manages this process by reversing the motor's operation, turning it into a generator.

Thermal management is also a critical feature integrated into motor controllers. Given the heat generated during high-power operations, motor controllers are equipped with systems to monitor and control the temperature of the motor and its components. Efficient thermal management ensures the motor operates within safe temperature limits, thereby extending its life and maintaining optimal performance.

Furthermore, motor controllers in IEVs are increasingly incorporating advanced connectivity features that enable real-time monitoring and diagnostics of the motor system. This connectivity allows for OTA updates and adjustments, which can improve the controller's algorithms and enhance the vehicle's performance based on aggregated data and new insights.

Major suppliers of motor controllers include Continental AG, Delphi Technologies, Sevcon, TI, and Kelly Controls.

Drivetrain Integration: Drivetrain integration is a complex and crucial aspect that defines how efficiently the vehicle utilizes its electric power to provide motion. The drivetrain in IEVs typically consists of the electric motor, the transmission system, power electronics, including the inverter and motor controller, and the battery pack. This integration is fundamental to optimizing vehicle performance, energy efficiency, and the driving experience.

At the heart of the drivetrain is the electric motor, which converts electrical energy from the battery into mechanical energy. The choice of motor (e.g., PMSM, induction motor) can vary based on the vehicle's specific performance, efficiency requirements, and cost constraints. Each motor type has its characteristics and integration requirements, which influence the overall design and functionality of the drivetrain.

The transmission system in IEVs is often simpler than that in ICE vehicles. Many IEVs use a single-speed transmission due to the electric motor's wide operating range, which can deliver high torque at low speeds and maintain power at high speeds. However, some newer models are exploring multispeed transmissions to enhance performance and efficiency at varying speeds.

Power electronics play a pivotal role in managing the flow of electrical energy within the drivetrain. The inverter, a key component, converts DC power from the battery to AC power needed by the electric motor. The motor controller regulates the motor's speed and torque based on inputs from the vehicle's control system, which responds to the driver's actions and road conditions.

Battery management is also integrated into the drivetrain system to ensure the battery operates within safe and efficient parameters. This integration involves monitoring and managing the charge state, temperature, and health of the battery to maximize its lifespan and performance. Furthermore, modern IEVs incorporate sophisticated software that controls and optimizes the interaction between these components. This software is capable of adjusting parameters in real time to improve efficiency, such as modifying the torque distribution in all-wheel-drive systems or managing regenerative braking to maximize energy recovery.

6.3. E/E Architectures

E/E architectures in vehicles refer to the integrated network of electronic systems and components that control various vehicle functions, and it is the core technology enabling electrification in IEVs [9, 10, 11, 12]. E/E architectures include engine management, transmission control, safety systems, infotainment, and, more recently, advanced driver-assistance systems (ADASs), and connectivity.

Before the 1970s, vehicles were primarily mechanical, with electronics playing a minimal role. Basic wiring systems were used to manage headlights, horns, and ignition systems. The 1970s marked a significant evolution with the introduction of microprocessors, which began to replace purely mechanical systems. The first notable use of electronics in vehicles was for engine control units that managed fuel injection and ignition timing. This period saw the beginning of the digital revolution in car manufacturing.

By the 1990s, the automotive industry witnessed a rapid increase in the number of ECUs within vehicles. Electronics began to control not only engines but also transmission, braking, and safety features such as airbags and anti-lock braking systems (ABSs). Then, during the 2000s, the focus shifted toward integration, where separate systems started to communicate and work together, leading to the development of more complex networked architectures.

The 2000s brought controller area network (CAN) and other networking technologies that allowed ECUs to share information, reducing wiring complexity and improving functionality. This era saw the proliferation of features such as electronic stability control and advanced infotainment systems.

The most recent developments in E/E architectures are characterized by a move toward centralization and higher integration, often described as domain controllers or zonal architectures. These are designed to handle the increased complexity and data volumes associated with ADAS, electrification, and autonomous driving technologies. The shift is toward fewer, more powerful central computers that manage various vehicle functions, aiming to improve efficiency, reduce costs, and enhance performance.

6.3.1. E/E Architectures for IEVs

E/E architectures in IEVs and traditional ICE vehicles exhibit distinct differences that reflect the divergent operational and environmental demands each faces. In IEVs, the trend is increasingly toward centralized architectures. This centralization involves using fewer, but more powerful, vehicle computers that control a broad array of functions, ranging from battery management to ADASs.

This shift is essential for handling the complex systems integrated into EVs, such as regenerative braking, electric power conversion, and sophisticated energy management systems. The centralized approach not only simplifies the architecture but also enhances the vehicle's ability to manage and utilize electrical energy efficiently. In contrast, traditional ICE vehicles typically feature a more distributed architecture. This setup includes multiple ECUs each dedicated to specific functions like engine control, transmission control, and other vehicular operations. Such a distribution results in a higher number of ECUs and leads to more segmented control across various vehicle systems, reflecting the mechanical nature of ICE vehicles and their reliance on optimizing fuel combustion and mechanical efficiency.

Furthermore, IEVs are designed to integrate seamlessly with renewable energy sources and smart grid technologies, incorporating capabilities for advanced energy flow management. This includes vehicle-to-grid (V2G) interactions and smart charging protocols, which are critical for enhancing energy efficiency and supporting the broader energy ecosystem. In contrast, ICE vehicles lack such capabilities primarily because their energy needs are satisfied through fuel combustion, and they do not interact with electrical energy sources.

Additionally, IEVs typically incorporate advanced connectivity and automation features necessitating robust E/E architectures capable of managing high-speed data transmissions. These features support functionalities like autonomous driving, remote vehicle diagnostics, and OTA updates, which are integral to the modern digital and automated landscape. Meanwhile, although modern ICE vehicles may include some level of connectivity and automated features, the complexity and integration level of these technologies are generally less advanced due to the inherently less complex nature of their E/E architectures.

Hence, the evolution toward more centralized, sophisticated, and integrated E/E architectures in EVs reflects the evolving needs of modern transportation. This evolution emphasizes efficiency, connectivity, and sustainability, aligning with the broader trends toward electrification and intelligent transportation systems.

6.3.2. Centralized E/E Architectures

Centralized E/E architectures integrate many vehicle functions, which were traditionally managed by multiple ECUs, into fewer, more potent central computing units (CCUs). The overarching aims of these architectures are to enhance vehicle performance, streamline software management, reduce hardware redundancy, and simplify the processes of updates and maintenance. Centralized E/E architectures are characterized by a few critical components:

- **CCUs:** These units are processors that consolidate the core computational tasks previously distributed among various ECUs. They oversee essential systems like propulsion, steering, and braking and process inputs from an array of sensors such as LiDAR, radar, and cameras. A key example of a CCU in commercial use is Tesla's FSD Computer, which processes data from an array of sensors to enable sophisticated driving features. Similarly, NVIDIA's DRIVE AGX offers a scalable and open computing platform that supports the autonomous vehicle ecosystem. It is designed to manage the extensive computational demands of autonomous driving, including deep learning and sensor fusion. The NXP S32G Vehicle Network Processor represents another approach, targeting service-oriented gateways that facilitate the transition to centralized E/E architectures. This processor is tailored for high-performance automotive computing, emphasizing improved networking, security, and data handling capabilities. Another significant product is the Qualcomm Snapdragon Ride Platform, which provides the necessary computing power for critical operations such as perception, planning, and actuation, playing a crucial role in supporting advanced, integrated vehicle architectures. Lastly, Mobileye's EyeQ5 focuses on delivering the processing power required for fully autonomous vehicles. It acts as a central hub for sensor data processing, vital for ADAS and autonomous driving functionalities, exemplifying the critical role of CCUs in modern vehicle systems.
- **Zone Controllers:** These controllers manage localized groups of functions such as lighting, window controls, and other minor ancillaries. While they handle routine actions locally to lessen the communication burden on the central system, they still communicate with the CCU for significant decisions. Examples of zone controllers in commercial vehicles include the following: Bosch's Vehicle Control Units manage specific vehicle domains like body electronics. These units are part of Bosch's broader strategy to enhance vehicle connectivity and automation by streamlining the

management of various subsystems through localized control units. Continental's Body High-Performance Computer (HPC) offers high-performance computers that act as zone controllers within a centralized E/E architecture. These systems manage functions in particular vehicle areas, such as the cockpit or body, centralizing the control for a group of related functions to improve efficiency and reduce system complexity. Visteon's SmartCore integrates infotainment, instrument clusters, and head-up displays into a single unit that manages multiple zones within the vehicle. This type of zone controller exemplifies the move toward more integrated systems that manage diverse functions through a single point of control to enhance user experience and system responsiveness.

- **High-Speed Data Networks:** These networks, often based on Ethernet, are crucial for managing the large data throughput that modern vehicles require, ensuring efficient data flow between sensors, actuaries, the CCU, and external networks. One prominent example of high-speed data networking is the Automotive Ethernet, which allows for high-speed data transmission between the vehicle's CCUs and other systems such as cameras, sensors, and infotainment units. Automotive Ethernet supports data speeds up to 10 Gbps, which is crucial for handling the vast amounts of data generated by modern vehicles. Another example is the CAN FlexRay, which is often used in conjunction with Ethernet. FlexRay provides a high-speed and reliable network backbone, suitable for safety-critical applications requiring determinism and fault tolerance, such as steering and braking systems. Additionally, Local Interconnect Network (LIN) is used for managing lower data rate components like seat controls, climate control, and lighting. Although not as fast as Ethernet or FlexRay, LIN is cost-effective and sufficient for less demanding applications.
- **Commercial Examples of Centralized E/E Architectures:** Several centralized E/E architectures have been commercialized and integrated into IEVs, such as the following: Tesla employs a highly centralized architecture with a custom-built chip designed specifically for autonomous driving, continuously enhanced through OTA software updates. Volkswagen Group's E3 2.0 Architecture supports the next generation of electric and autonomous vehicles with significantly fewer ECUs, centralizing control in high-capacity vehicle computers. General Motors' Vehicle Intelligence Platform (VIP) supports advanced technologies like vehicle-to-everything (V2X) communications and enhanced cybersecurity measures, promoting flexibility and scalability across GM's global vehicle lineup. Mercedes-Benz MB.OS is a centralized operating system to manage everything from infotainment to autonomous driving functions across its fleet, aiming to simplify software updates and improve service integration.

6.4. Skateboard Chassis

The skateboard chassis represents a transformative design in the architecture of IEVs, known for its simplicity and modularity for integrating different electrification technologies [13, 14]. The skateboard chassis consists of a flat, long platform which houses all the major components such as the battery pack, electric motors, and suspension system. This design not only allows for the battery to be placed on the floor of the vehicle, which lowers the center of gravity and improves handling, but it also simplifies manufacturing processes (Figure 6.1).

Figure 6.1 The skateboard chassis.

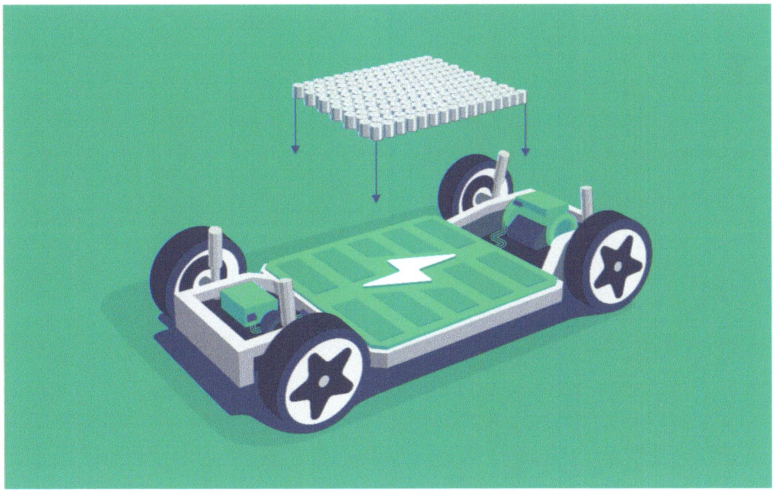

© Petovarga.

One of the primary advantages of the skateboard chassis over traditional vehicle architectures is its versatility. It can support various body styles and vehicle types on the same basic structure. This flexibility allows manufacturers to produce a range of vehicles—from sedans to sport utility vehicles (SUVs)—on a single platform, reducing development time and costs. It also supports easier upgrades and maintenance due to its modular nature, and components can be replaced or upgraded individually without redesigning the entire vehicle.

The skateboard design provides significant benefits in terms of interior space. Without the need for a traditional engine bay and drivetrain, it offers more cabin space, enabling a more spacious interior layout. This feature is particularly advantageous for EVs, as it allows designers to rethink passenger and cargo space without the constraints imposed by conventional vehicle components. Moreover, the skateboard platform is conducive to integrating advanced technologies. For instance, the use of drive-by-wire technology replaces mechanical linkages with electronic controls, offering new possibilities for vehicle interiors and control systems. This integration can lead to more personalized and comfortable driving experiences, as the car can be tailored more closely to the driver's preferences and needs.

Skateboard chassis has been adopted by major automotive manufacturers like Tesla, VW, and Hyundai—each with their own unique versions of the platform. Tesla's platform underpins models such as the Model S, Model X, and Model 3. The skateboard design allows for a flat floor in the vehicle's interior, enhancing space and battery integration. Volkswagen Group's Modular electric drive matrix (MEB) platform has been used across various brands within the group. This platform supports a range of EVs with different body styles, promoting design flexibility and cost-efficiency. Hyundai's Electric Global Modular Platform (E-GMP) is specifically developed for EVs. This platform is used in Hyundai and Kia EVs, enabling rapid model development and enhanced electric performance. Rivian's skateboard platform integrates a quad-motor system, offering high performance and off-road capabilities. General Motors' Ultium Platform empowers models like the GMC Hummer EV and Cadillac Lyriq, focusing on versatility and advanced electric propulsion.

6.5. Summary

Electrification technologies have emerged as a transformative force in the automotive industry, driving a shift from traditional fossil fuel reliance to more sustainable, efficient EVs (IEVs). This chapter explores the various facets of electrification technologies which include battery systems, electric motors, power electronics, and advanced E/E architectures. These components are crucial for enhancing vehicle performance, reducing emissions, and improving the overall sustainability of automotive transport.

References

1. Zhang, R., Xia, B., Li, B., Cao, L. et al., "State of the Art of Lithium-Ion Battery SOC Estimation for Electrical Vehicles," *Energies* 11, no. 7 (2018): 1820.
2. Xiong, R. and Shen, W., *Advanced Battery Management Technologies for Electric Vehicles* (Hoboken: John Wiley & Sons, 2019).
3. Kim, J., Oh, J., and Lee, H., "Review on Battery Thermal Management System for Electric Vehicles," *Applied Thermal Engineering* 149 (2019): 192-212.
4. Sissonarchive, P., "How Does an EV Battery Actually Work?" *MIT Technology Review*, accessed May 5, 2024, https://www.technologyreview.com/2023/02/17/1068037/how-do-ev-batteries-work/.
5. Habib, A.A., Hasan, M.K., Issa, G.F., Singh, D. et al., "Lithium-Ion Battery Management System for Electric Vehicles: Constraints, Challenges, and Recommendations," *Batteries* 9, no. 3 (2023): 152.
6. Chan, C.C. and Chau, K.T., "An Overview of Power Electronics in Electric Vehicles," *IEEE Transactions on Industrial Electronics* 44, no. 1 (1997): 3-13.
7. Habib, S., Khan, M.M., Abbas, F., Ali, A. et al., "Contemporary Trends in Power Electronics Converters or Charging Solutions of Electric Vehicles," *CSEE Journal of Power and Energy Systems* 6, no. 4 (2020): 911-929.
8. Emadi, A., Lee, Y.J., and Rajashekara, K., "Power Electronics and Motor Drives in Electric, Hybrid Electric, and Plug-In Hybrid Electric Vehicles," *IEEE Transactions on Industrial Electronics* 55, no. 6 (2008): 2237-2245.
9. Ribbens, W.B. and Mansour, N.P., *Understanding Automotive Electronics*. Vol. 350 (Amsterdam, the Netherlands: Newnes, 2003).
10. Leen, G. and Heffernan, D., "Expanding Automotive Electronic Systems," *Computer* 35, no. 1 (2002): 88-93.
11. Bandur, V., Selim, G., Pantelic, V., and Lawford, M., "Making the Case for Centralized Automotive E/E Architectures," *IEEE Transactions on Vehicular Technology* 70, no. 2 (2021): 1230-1245.
12. Haeberle, M., Heimgaertner, F., Loehr, H., Nayak, N. et al., "Softwarization of Automotive E/E Architectures: A Software-Defined Networking Approach," in *2020 IEEE Vehicular Networking Conference (VNC)*, New York, 2020, 1-8, IEEE.
13. Pinheiro, L., Passarinho, F., Teixeira, A., Silva, H. et al., "New Skate Chassis Concept for Electric Vehicles," SAE Technical Paper 2012-01-0039 (2012), doi:https://doi.org/10.4271/2012-01-0039.
14. Sharath Chandra, N. and Vikas, R., "Design and Analysis of Skateboard Chassis," in: Krishna, V., Seetharamu, K.N., and Joshi, Y.K. (eds), *Recent Advances in Hybrid and Electric Automotive Technologies: Select Proceedings of HEAT 2021* (Singapore: Springer Nature Singapore, 2022), 263-276.

Chapter 07

Enabling Technologies—Autonomous Driving

Autonomous driving is not merely an evolution in transportation; it is a revolution in human mobility, redefining our concepts of driving, safety, and time management on the roads.

—Anonymous

In this chapter, we provide a concise overview of autonomous driving technologies. Initially, we examine the technology stack that forms the foundation of autonomous driving systems. Following this, we explore in detail the algorithms that drive autonomous decision-making, highlighting the role of edge computing in facilitating real-time processing capabilities, and the integration of cloud computing for enhanced data analysis and storage solutions. Additionally, we address the scalability challenges faced by current autonomous driving methodologies, underscoring the need for adaptable and efficient frameworks to support the evolving demands of autonomous mobility. For an in-depth exploration of autonomous driving technologies and their applications, readers are encouraged to consult References [1, 2, 3, 4, 5].

7.1. Market Landscape

Autonomous driving technologies, including assisted driving technologies, also known as ADASs, and fully autonomous driving technologies (robotaxi), have significantly evolved in recent years. In the robotaxi market, several companies are making significant progress: Waymo has been a pioneer in this field, operating self-driving fleets in cities like Phoenix and San Francisco. Waymo utilizes a robust set of sensors and proprietary software to navigate complex driving environments. Cruise focuses on developing and deploying autonomous vehicles in urban settings, leveraging GM's vast automotive experience with innovative autonomous technology to pilot driverless taxis. Uber, through its partnership with Motional, uses Hyundai IONIQ 5 vehicles equipped with advanced self-driving hardware to offer autonomous ride-hailing services in select cities, indicating a shift toward commercial viability in the ride-sharing sector.

In the ADAS market, key players such as Tesla, NVIDIA, XPeng, MobilEye, Huawei, Bosch, and Continental are at the forefront of this transformation, each contributing unique innovations that enhance vehicle intelligence and safety. Key ADAS product features are summarized as follows:

- Adaptive cruise control (ACC): ACC automatically adjusts the vehicle's speed to maintain a safe distance from the car ahead, making long journeys on the highway more comfortable and safer.
- Blind spot monitoring (BSM): Sensors detect other vehicles in the driver's blind spots and alert the driver through visual or auditory signals, helping to prevent potential side collisions during lane changes.
- Pedestrian detection: This feature uses cameras and sensors to identify pedestrians near the vehicle and can automatically apply brakes if a collision is imminent, enhancing safety in urban environments.
- Driver monitoring systems (DMSs): These systems monitor the driver's behavior to detect signs of distraction or fatigue, providing alerts to help maintain focus on the road.
- Traffic sign recognition (TSR): Cameras recognize traffic signs and provide real-time information to the driver, ensuring compliance with traffic laws such as speed limits and no-entry signs.
- Automatic parking: This feature assists drivers in parking maneuvers by automatically steering the vehicle into a parking spot, often requiring the driver only to control the brakes and accelerator.
- Emergency braking: This system detects potential collisions and automatically applies the brakes to reduce impact or avoid the collision altogether.
- Lane keeping assist (LKA): This feature helps prevent unintentional lane departures by alerting the driver or actively steering the vehicle back into the lane.

In the ADAS market, Tesla continues to refine its Autopilot system, which includes features like automatic lane changes and adaptive cruise control. Despite being classified as Level 2 autonomy, Tesla's system is one of the most advanced on the market, capable of semiautonomous driving under specific conditions.

NVIDIA has made significant strides with its DRIVE platform, which is not only powering vehicles from various manufacturers but also pushing the boundaries of what assisted driving systems can do. The platform's high-performance computational capabilities enable sophisticated sensor processing and deep learning applications essential for next-generation ADAS.

XPeng has integrated advanced driving features in its vehicles, particularly through the XPILOT system, which offers navigational guided pilot and valet parking, showcasing substantial advancements in both hardware integration and software algorithms.

Huawei has also entered the automotive scene with its HI solutions, focusing on smart cockpits and intelligent driving systems. Huawei leverages its strong background in telecommunications and AI to enhance connectivity and automate driving tasks.

Mobileye's ADAS offerings center around its proprietary EyeQ® family of SoC designs, which are integral to processing inputs from multiple cameras mounted on the vehicle. One of Mobileye's hallmark features is its collision avoidance technology, which utilizes an intricate process of real-time image recognition to detect potential hazards on the road, such as vehicles, pedestrians, cyclists, and lane markings. This system provides drivers with audio and visual warnings to prevent or mitigate collisions. Mobileye's Lane Departure Warning (LDW) system uses sophisticated algorithms to recognize lane markings and boundaries. If the system detects that the vehicle is drifting out of its lane without an activated turn signal, it alerts the driver, significantly reducing the risk of accidents caused by lane drifting.

Bosch is one of the top players in the ADAS market, renowned for its wide array of driver assistance technologies. Bosch has been pivotal in advancing sensor technologies and software solutions that underpin many ADAS features. Their systems are widely used across the automotive industry, given Bosch's longstanding reputation for quality and innovation in automotive components.

Continental is another major supplier, known for integrating advanced radar and camera systems to facilitate complex ADAS functionalities. Continental's solutions are designed to enhance vehicle intelligence and road safety, supporting automakers in their push toward more autonomous driving features.

The ADAS market is driven by several trends, including the increasing demand for vehicle safety, the push toward autonomous vehicles, and regulatory mandates requiring vehicle safety features. The integration of ADAS technologies is becoming a standard across IEVs, influenced by consumer preferences for safer, more comfortable, and technologically advanced driving options. In terms of market share, companies like Bosch and Continental continue to lead due to their long-standing industry presence and comprehensive technology offerings. However, newer players like Mobileye and Huawei are rapidly expanding their market presence by leveraging their expertise in high-tech sectors and integrating innovative technologies into their ADAS solutions.

7.2. Autonomous Driving Technologies Overview

Figure 7.1 illustrates that autonomous driving technology is not a singular entity but a complex integration of multiple subsystems. This can be broadly categorized into three primary components:

- Algorithms: This encompasses aspects such as sensing, perception, and decision-making, which are essential for navigating intricate situations [3, 4].
- Edge systems: This includes both the operating system and the hardware platform. These systems are crucial in synthesizing the algorithms to fulfill real-time performance and reliability demands. For instance, if a camera operates at 60 Hz, the edge systems must ensure that the most time-consuming process in the pipeline is completed in less than 16 ms [6–10].

Figure 7.1 Autonomous driving technology stack.

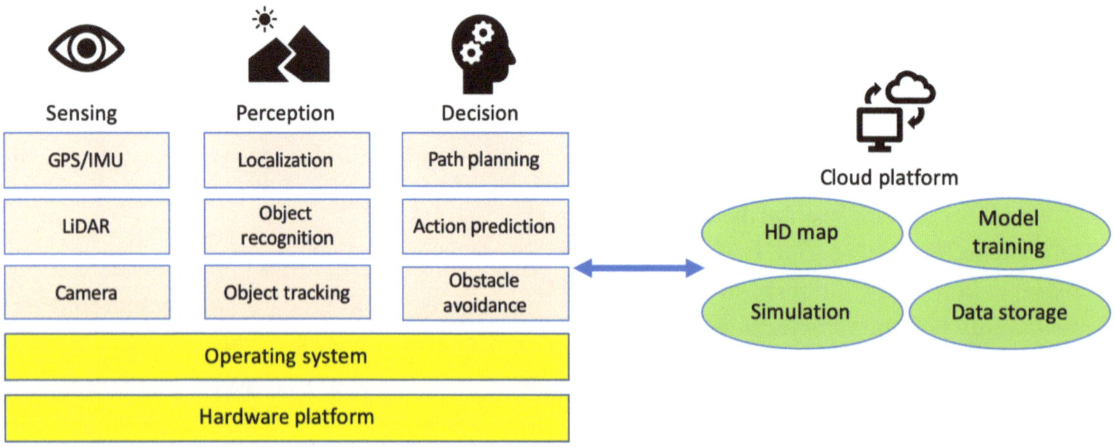

- Cloud platform: This provides essential offline computing and storage facilities for autonomous vehicles. The cloud platform is instrumental in testing and refining new algorithms, updating HD maps, and enhancing models for recognition, tracking, and decision-making [11–15].

In essence, the algorithm subsystem is responsible for interpreting raw sensor data to comprehend the vehicle's surroundings and determine appropriate actions. The edge systems integrate these algorithms to adhere to stringent timing and reliability standards. Meanwhile, the cloud platform offers the computational and storage resources necessary for the ongoing development and improvement of autonomous driving technologies.

7.3. Autonomous Driving Algorithms

The algorithms component of autonomous driving technology is a critical pillar, comprising three key functions: sensing involves the extraction of meaningful information from the raw data captured by various sensors. It is the foundational step in processing the external data that the vehicle receives. Perception involves localizing the vehicle, determining its precise position in relation to its surroundings, as well as understanding the current environment and interpreting the data gathered to form a comprehensive picture of the immediate external conditions. Decision-making is the process of determining the most appropriate actions for the vehicle to take in order to navigate to its destination safely and reliably. It involves making calculated choices based on the information processed in the sensing and perception stages, ensuring the vehicle's actions are both safe and efficient in reaching its target destination.

7.3.1. Sensing

An autonomous car is equipped with a variety of major sensors, each offering unique advantages and limitations. To enhance reliability and safety, data from these multiple sensors are integrated. The key sensors include the following:

- GPS/IMU (inertial measurement unit): The GPS/IMU system is pivotal for vehicle localization. The GPS provides a global position estimate, though it updates slowly at around 10 Hz. On the other hand, the IMU, despite its errors accumulating over time, offers more frequent updates at or above 200 Hz. By combining GPS and IMU data, the system can provide accurate and real-time localization updates.
- LiDAR: LiDAR sensors are crucial for mapping, localization, and obstacle avoidance. They operate by emitting light beams and measuring the time taken for the reflections to return, thus determining distances with high accuracy. A typical LiDAR unit, like the Velodyne 64-beam laser, rotates at 10 Hz and can take approximately 1.3 million readings per second. This makes LiDAR highly effective for producing HD maps, localizing moving vehicles, and detecting obstacles.
- Cameras: Cameras play a significant role in object recognition and tracking, aiding in functions like lane detection, traffic light recognition, and pedestrian monitoring. Modern autonomous vehicles often mount eight or more 1080 p cameras to cover all around the vehicle. Operating at 60 Hz, these cameras collectively generate about 1.8 GB of raw data per second.
- Radar and Sonar: These systems are primarily used for close-range obstacle avoidance, acting as a last line of defense. Radar and sonar data, which provide information on distance and velocity relative to objects in the vehicle's path, are crucial in emergency situations like imminent collisions. These data are typically fed directly to the control processor for immediate response actions such as braking, swerving, or seat belt pretensioning, bypassing the main computation pipeline due to its urgency.

The integration of data from these diverse sensors is key to the functionality of autonomous vehicles, ensuring they operate safely and efficiently in various driving conditions.

7.3.2. Perception

The sensor data are then fed into the perception stage to provide an understanding of the vehicle's environment. The three main tasks in autonomous driving perception are localization, object detection, and object tracking. GPS/IMU can be used for localization, and, as mentioned above, GPS provides fairly accurate localization results but with a comparatively low update rate, while an IMU provides very fast updates at a cost of less accurate results.

Localization: We can thus use Kalman filter techniques to combine the advantages of the two and provide accurate and real-time position updates. As shown in Figure 7.2, it works as follows: the IMU updates the vehicle's position every 5 ms, but the error accumulates with time. Fortunately, every 100 ms, a GPS update is received, which helps correct the IMU error. By running this propagation and update model, the GPS/IMU combination can generate fast and accurate localization results.

Figure 7.2 Autonomous driving processing pipeline.

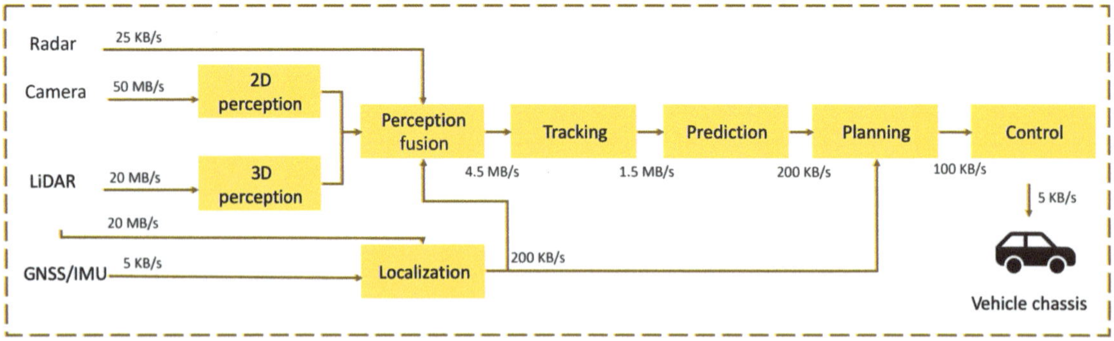

Nonetheless, we cannot solely rely on this combination for localization for three reasons: (1) The accuracy is only about one meter; (2) the GPS signal has multipath problems, meaning that the signal may bounce off buildings, introducing more noise; and (3) GPS requires an unobstructed view of the sky and would thus not work in environments such as tunnels.

Vision-based localization can be implemented as the following simplified pipeline: (1) By triangulating stereo image pairs, we first obtain a disparity map which can be used to derive depth information for each point; (2) by matching salient features between successive stereo image frames, we can establish correlations between feature points in different frames. We could then estimate the motion between the past two frames; and (3) by comparing the salient features against those in the known map, we could also derive the current position of the vehicle. However, such a vision-based localization approach is very sensitive to lighting conditions and, thus, this approach alone would not be reliable.

Based on the observation that inertial sensors are prone to drift and wheel odometry is unreliable in bumpy "off-road" terrain, we can utilize visual odometry for real-time localization. Stereo visual odometry works by estimating frame-to-frame camera motion from successive stereo image pairs, and it has been successfully applied to autonomous driving. For instance, one flavor of visual odometry algorithms works in the following way: (1) It makes no prior assumptions about camera motion, and (2) it operates on dense disparity images computed by a separate stereo algorithm. The authors have demonstrated that after 4000 frames and 400 m of travel, position errors are typically less than 1 m (0.25% of distance traveled); and the processing time is only about 20 ms on a 512 × 384 image. The basic algorithm works as follows: For a given pair of frames, (1) detect features in each frame with corner feature detection; (2) match features between frames using sum-of-absolute differences over local windows; (3) find the largest set of self-consistent matches (inlier detection); and (4) find the frame-to-frame motion that minimizes the reprojection error for features in the inlier set. The feature matching stage inevitably produces some incorrect correspondences, which, if left intact, will unfavorably bias the frame-to-frame motion estimate. A common solution to this problem is to use a robust estimator that can tolerate some number of false matches. The core intuition is that the 3-D locations of features must obey a rigidity constraint and that this constraint can be used to identify sets of features that are mutually consistent prior to computing the frame-to-frame motion estimate. Note that this algorithm does not require an initial motion estimate and therefore can handle very large image translations.

Object Recognition and Tracking: Originally LiDAR was used mostly to perform object detection and tracking tasks in autonomous vehicles, since LiDAR provides very accurate depth information. In recent years, however, we have seen the rapid development of deep learning technology, which achieves significant object detection and tracking accuracy. Convolution neural network (CNN) is a type of deep neural network that is widely used in object recognition tasks. A general CNN evaluation pipeline usually consists of the following layers. (1) The convolution layer contains different filters to extract different features from the input image. Each filter contains a set of "learnable" parameters that will be derived after the training stage. (2) The activation layer decides whether to activate the target neuron or not. (3) The pooling layer reduces the spatial size of the representation to reduce the number of parameters and consequently the computation in the network. (4) In the fully connected layer, neurons have full connections to all activations in the previous layer.

Object tracking refers to the automatic estimation of the trajectory of an object as it moves. After the object to track is identified using object recognition techniques, the goal of object tracking is to automatically track the trajectory of the object subsequently. This technology can be used to track nearby moving vehicles as well as people crossing the road to ensure that the current vehicle does not collide with these moving objects. In recent years, deep learning techniques have demonstrated advantages in object tracking compared to conventional computer vision techniques. Specifically, by using auxiliary natural images, a stacked autoencoder can be trained offline to learn generic image features that are more robust against variations in viewpoints and vehicle positions. Then, the offline trained model can be applied for online tracking.

7.3.3. Action

Based on the understanding of the vehicle's environment, the decision stage can generate a safe and efficient action plan in real time. One of the main challenges for human drivers when navigating through traffic is to cope with the possible actions of other drivers which directly influence their own driving strategy. This is especially true when there are multiple lanes on the road or when the vehicle is at a traffic change point. To make sure that the vehicle travels safely in these environments, the decision unit generates predictions of nearby vehicles, and decides on an action plan based on these predictions. To predict actions of other vehicles, one can generate a stochastic model of the reachable position sets of the other traffic participants and associate these reachable sets with probability distributions.

- **Path Planning:** Planning the path of an autonomous, agile vehicle in a dynamic environment is a very complex problem, especially when the vehicle is required to use its full maneuvering capabilities. A brute force approach would be to search all possible paths and utilize a cost function to identify the best path. However, the brute force approach would require enormous computation resources and may be unable to deliver navigation plans in real time. In order to circumvent the computational complexity of deterministic, complete algorithms, probabilistic planners have been utilized to provide effective real-time path planning.

- **Obstacle Avoidance:** As safety is the paramount concern in autonomous driving, we usually employ at least two levels of obstacle avoidance mechanisms to ensure that the vehicle does not collide with obstacles. The first level is proactive and is based on traffic predictions. At runtime, the traffic prediction mechanism generates measures like time to collision or predicted minimum distance, and based on this information, the obstacle avoidance mechanism is triggered to perform local path replanning. If the proactive mechanism fails, the second level, the reactive mechanism, using radar data, would take over. Once the radar detects an obstacle ahead of the path, it would override the current control to avoid the obstacles.

7.4. Autonomous Driving Edge Systems

The edge systems integrate the above-mentioned algorithms together to meet real-time and reliability requirements. Some of the challenges are as follows: The system needs to ensure that the processing pipeline is fast enough to process the enormous amount of sensor data generated; if a part of the system fails, it must be sufficiently robust to recover from the failure; and, in addition, it needs to perform all the computation under strict energy and resource constraints.

7.4.1. Middleware

Robot Operating System (ROS) is a powerful distributed computing framework tailored for robotics applications, and it has been widely used. Each robotic task (such as localization) is hosted in a ROS node. ROS nodes can communicate with each other through topics and services. It is a great operating system for autonomous driving except that it suffers from several problems: (1) Reliability: ROS has a single master and no monitor to recover failed nodes; (2) performance: when sending out broadcast messages, it duplicates the message multiple times, leading to performance degradation; and (3) security: it has no authentication and encryption mechanisms. Although ROS 2.0 promised to fix these problems, ROS 2.0 itself has not been extensively tested, and many features are not yet available. Therefore, in order to use ROS in autonomous driving, we need to solve these problems first.

7.4.2. Hardware Platform

To understand the challenges in designing a hardware platform for autonomous driving, let us examine the computing platform implementation from a leading autonomous driving company. It consists of two compute boxes, each equipped with an Intel Xeon E5 processor and four to eight Nvidia K80 GPU accelerators. The second compute box performs exactly the same tasks and is used for reliability: in case the first box fails, the second box can immediately take over. In the worst case, when both boxes run at their peak, this would mean over 5000 W of power consumption, which would consequently generate an enormous amount of heat. Also, each box costs \$20,000–\$30,000, making the whole solution unaffordable to average consumers.

The power, heat dissipation, and cost requirements of this design prevent autonomous driving from reaching the general public. To explore the edges of the envelope and understand how well an autonomous driving system could perform on an ARM mobile SoC, we implemented a simplified, vision-based autonomous driving system on an ARM-based mobile SoC with a peak power consumption of 15 W. As it turns out, the performance was close to our requirements: the localization pipeline was capable of processing 25 images per second, almost keeping up with an image generation rate of 30 images per second. The deep learning pipeline was capable of performing 2–3 object recognition tasks per second. The planning and control pipeline could plan a path within 6 ms. With this system, we were able to drive the vehicle at around 5 mph without any loss of localization.

7.5. Autonomous Driving Cloud Platform

Autonomous vehicles are mobile systems, and therefore, they need a cloud platform to provide support. The two main functions provided by the cloud include distributed computing and distributed storage. It has several applications, including simulation, which is used to verify new algorithms; HD map production; and deep learning model training.

To build such a platform, in one commercial instantiation [11], we used Spark for distributed computing, OpenCL for heterogeneous computing, and Alluxio for in-memory storage. We have managed to deliver a reliable, low-latency, and high-throughput autonomous driving cloud by integrating Spark, OpenCL, and Alluxio together.

The first application of such a system is simulation [14, 15]. When we develop a new algorithm, we need to test it thoroughly before we can deploy it on cars. If we were to test it on real cars, the cost would be enormous, and the turnaround time would be too long. Therefore, we usually test it on simulators, such as replaying data through ROS nodes. However, if we were to test the new algorithm on a single machine, either it is going to take too long, or we do not have enough test coverage.

To solve this problem, we have developed a distributed simulation platform such that we use Spark to manage distributed computing nodes, and on each node, we run a ROS replay instance. In an autonomous driving object recognition test set that we used, it took 3 h to run on a single server; by using the distributed system we developed, the test finished within 25 min when we scaled to eight machines.

HD map production is a complex process that involves many stages, including raw data processing, point cloud production, point cloud alignment, 2-D reflectance map generation, HD map labeling, as well as the final map generation. Using Spark, we connected all these stages together in one Spark job. More importantly, Spark provides an in-memory computing mechanism, such that we do not have to store the intermediate data on a hard disk, thus greatly reducing the performance of the map production process [16–19].

As we use different deep learning models in autonomous driving, it is imperative to provide updates to continuously improve the effectiveness and efficiency of these models. However, since the amount of raw data generated is enormous, we would not be able to achieve fast model training using single servers.

7.6. Autonomous Driving 2.0

Current commercial autonomous driving systems have shown their age and become difficult to scale after more than a decade of intense industry efforts to improve and adapt [5]. Over the past five years, although an enormous amount of investment has been poured into autonomous driving, we did not observe significant growth in the number of vehicles under operation, which increased only from 400 in 2018 to 1500 in 2022. The operation mileage per year increased only from 2 to 5 million miles. Most importantly, there are still over 2000 disengagement incidents per year. Given this trend, we are still years away from serious commercial operations of autonomous vehicles.

As reviewed above, existing autonomous driving architecture (dubbed Autonomy 1.0) is modular and consists of functional modules such as sensing, perception, localization, HD maps, prediction, planning, and control, each further consisting of several functional submodules integrated by explicit and handcrafted logic. Most decision-making tasks, such as planning, which is responsible for generating optimal and drivable paths, are solved with constraint optimization under a set of hand-tuned rules.

When a disengagement incident happens, engineers usually have to go through a long process of debugging to identify which specific module or rule may have been the root cause of the disengagement, and then optimize that module or develop logic changes to handle the specific problem. Often, due to intricate dependency and coupling among modules or rules, the new software version leads to other problems that need to be addressed, thus greatly slowing down the development process.

The Autonomy 1.0 software stack over time became a complicated collection of ad hoc rules and a set of interdependent modules for handling various long-tail events, which has been increasingly difficult to debug, maintain, and evolve for improved performance. Taking the open-source project Apollo as an example, its perception module alone consists of multiple individual leaning-based submodules to accomplish object detection in 2-D images, LiDAR point cloud segmentation, traffic light detection, lane detection, and others. To integrate information from these perception submodules, a postprocessing module then fuses 2-D and 3-D information and outputs an integrated representation of the environment to the downstream prediction module. The planning module makes decisions and plans routes based on the data from the prediction, localization, and map modules. These modules often have strong dependencies among themselves. Making changes to one module not only impacts the overall system performance, possibly violating real-time constraints and resource allocation, but also impacts the algorithmic performance of other downstream modules due to the distributional shift of data. The whole system has become complicated and even brittle, demanding an enormous amount of engineering resources to maintain, let alone to scale.

We summarize the three Autonomy 1.0's major scalability bottlenecks below. First, the design of Autonomy 1.0 systems demands extensive engineering efforts to define software interfaces, distribute data among modules, and map various workloads in a heterogeneous computing system. It is challenging, given the complexity, to debug and continuously update the software stack. The myriad of components also makes it challenging to schedule tasks and optimize the latency of the unwieldy stack at run-time. As a result, typical Autonomy 1.0 systems exhibit large latency variations [20], which can harm the reliability of the autonomous driving system.

Second, Autonomy 1.0 systems depend on fleets of physical vehicles operated by humans to collect data and perform system-level tests. This is a time-consuming and expensive process that is difficult to scale out. The scalability issue will only get worse as increasingly more modules of autonomy stack adopt data-driven approaches, which requires continuous collection and labeling, because any specific instance of the recorded data reflects only a particular subset of the world states.

Third, Autonomy 1.0 systems consist of rule-based processing logic and hand-crafted interfaces, which makes them difficult to generalize to new environments. This is because the complexity and diversity of real-world environments make it difficult to design the autonomy system to anticipate all possible challenging scenarios, whether for perception or planning. As a result, Autonomy 1.0 systems are often overfitted to frequently operated regions and common situations. To handle new environments and newly encountered rare cases, additional changes to the system are required, which is increasingly difficult and time-consuming.

Recent research breakthroughs in AI, such as Transformer, LLM, and offline reinforcement learning, have sparked new ideas in architecture design, data and model infrastructure, and engineering practices of autonomous driving, leading to a new development paradigm, which we dub Autonomy 2.0. The key of Autonomy 2.0 is scalability, which is delivered through two ingredients: (1) a software stack that improves continuously with increasing scale of data and compute resources; and (2) a simulation paradigm based on digital twins for algorithmic exploration using large-scale, real-time, realistic data before deployment.

Any autonomous machine performs two main tasks: perception and action, reflecting the natural dichotomy of the past and the future. The perception task observes the environment and infers its current state based on observations so far. The action task, based on these observations, chooses an appropriate sequence of actions to achieve goals while considering how the environment may evolve in the near future. The software stack in Autonomy 2.0, thus, naturally consists of a perception module and an action module. Unlike in Autonomy 1.0 where each module is implemented by a number of submodules, there is strong evidence that the two modules, in Autonomy 2.0, will each be implemented as a single large deep learning model, likely based on Transformers or its variants due to their ability to generalize, as demonstrated in their recent successes in LLMs.

Before describing how the two-model architecture will look like in Autonomy 2.0, we will first discuss why such an architectural design choice is key to scalability. The two-model architecture addresses the complexity bottleneck by drastically reducing the amount of code that needs to be maintained and reasoned about. We compare the lines of code in the Apollo Perception module, which represents the

Autonomy 1.0 approach, with an example of the perception module in Autonomy 2.0, BEVFormer [21]. The Apollo Perception module's size is ten times larger than BEVFormer, and BEVFormer has achieved state-of-the-art perception results. The software architecture also handles corner cases through data-driven model learning instead of hand-crafted logic, and thus addresses the generalization bottleneck in Autonomy 1.0.

7.7. Summary

Autonomous driving is essential to IEVs for several reasons, primarily enhancing safety, efficiency, and environmental sustainability. One of the fundamental advantages of autonomous driving in IEVs is the potential to significantly reduce traffic accidents caused by human error. In this chapter, we review the market landscape of autonomous driving technologies, and then we delve into the technical details of various components of the autonomous driving technology stack. For readers who are interested in various application scenarios, please refer to [22, 23]. For readers who are interested in methodologies of teaching autonomous driving technologies, please refer to [24, 25].

References

1. Liu, S. and Gaudiot, J.L., "Rise of the Autonomous Machines," *Computer* 55, no. 1 (2022): 64-73.
2. Liu, S., Peng, J., and Gaudiot, J.L., "Computer, Drive My Car!" *Computer* 50, no. 1 (2017): 8-8.
3. Liu, S., Li, L., Tang, J., Wu, S. et al., *Creating Autonomous Vehicle Systems* (San Rafael, CA: Morgan & Claypool, 2018).
4. Liu, S., *Engineering Autonomous Vehicles and Robots: The DragonFly Modular-Based Approach* (Hoboken: John Wiley & Sons, 2020).
5. Wu, S., Yu, B., Liu, S. and Zhu, Y., "Autonomy 2.0: The Quest for Economies of Scale," arXiv preprint arXiv:2307.03973, 2023.
6. Liu, S., Liu, L., Tang, J., Yu, B. et al., "Edge Computing for Autonomous Driving: Opportunities and Challenges," *Proceedings of the IEEE* 107, no. 8 (2019): 1697-1716.
7. Liu, S., Tang, J., Zhang, Z., and Gaudiot, J.L., "Computer Architectures for Autonomous Driving," *Computer* 50, no. 8 (2017): 18-25.
8. Tang, J., Liu, S., Liu, L., Yu, B. et al., "LoPECS: A Low-Power Edge Computing System for Real-Time Autonomous Driving Services," *IEEE Access* 8 (2020): 30467-30479.
9. Tang, J., Yu, R., Liu, S., and Gaudiot, J.L., "A Container Based Edge Offloading Framework for Autonomous Driving," *IEEE Access* 8 (2020): 33713-33726.
10. Jiang, X., Luo, X., Guan, N., Dong, Z. et al., "Analysis and Optimization of Worst-Case Time Disparity in Cause-Effect Chains," in *2023 Design, Automation & Test in Europe Conference & Exhibition (DATE)*, Antwerp, Belgium, 2023, 1-6, IEEE.
11. Liu, S., Tang, J., Wang, C., Wang, Q. et al., "A Unified Cloud Platform for Autonomous Driving," *Computer* 50, no. 12 (2017): 42-49.
12. Tang, J., Liu, S., Cao, J., Sun, D. et al., "π-Hub: Large-Scale Video Learning, Storage, and Retrieval on Heterogeneous Hardware Platforms," *Future Generation Computer Systems* 102 (2020): 514-523.
13. Liu, L., Liu, S., and Shi, W., "4C: A Computation, Communication, and Control Co-Design Framework for CAVs," *IEEE Wireless Communications* 28, no. 4 (2021): 42-48.
14. Yu, B., Chen, C., Tang, J., Liu, S. et al., "Autonomous Vehicles Digital Twin: A Practical Paradigm for Autonomous Driving System Development," *Computer* 55, no. 9 (2022): 26-34.
15. Yu, B., Tang, J., and Liu, S.S., "Autonomous Driving Digital Twin Empowered Design Automation: An Industry Perspective," in *2023 60th ACM/IEEE Design Automation Conference (DAC)*, San Francisco, 2023, 1-4, IEEE.
16. Jiao, J., "Machine Learning Assisted High-Definition Map Creation," in *2018 IEEE 42nd Annual Computer Software and Applications Conference (COMPSAC)*, Vol. 1, Tokyo, Japan, 2018, 367-373, IEEE.

17. Li, D., Tang, J., and Liu, S., "Brief Industry Paper: An Edge-Based High-Definition Map Crowdsourcing Task Distribution Framework for Autonomous Driving," in *2021 IEEE 27th Real-Time and Embedded Technology and Applications Symposium (RTAS)*, Nashville, 2021, 453-456, IEEE.

18. Tang, J. and Liu, S., "Edge High-Precise Map Service for Autonomous Driving," *ZTE Technology Journal* 25, no. 3 (2019).

19. Yang, Z., Yu, B., Hu, W., Tang, J. et al., "π-Map: A Decision-Based Sensor Fusion with Global Optimization for Indoor Mapping," in *2020 IEEE/RSJ International Conference on Intelligent Robots and Systems (IROS)*, Las Vegas, 2020, 4821-4827, IEEE.

20. Yu, B., Hu, W., Xu, L., Tang, J. et al., "Building the Computing System for Autonomous Micromobility Vehicles: Design Constraints and Architectural Optimizations," in *2020 53rd Annual IEEE/ACM International Symposium on Microarchitecture (MICRO)*, Athens, Greece, October 2020, 1067-1081, IEEE.

21. Li, Z., Wang, W., Li, H., Xie, E. et al., "BEVFormer: Learning Bird's-Eye-View Representation from Multi-Camera Images via Spatiotemporal Transformers," in: Avidan, S., Brostow, G., Cissé, M., Farinella, G.M. et al. (eds), *European Conference on Computer Vision* (Cham, Switzerland: Springer Nature Switzerland, 2022), 1-18.

22. Li, B., Liu, S., Tang, J., Gaudiot, J.L. et al., "Autonomous Last-Mile Delivery Vehicles in Complex Traffic Environments," *Computer* 53, no. 11 (2020): 26-35.

23. Liu, S. and Gaudiot, J.L., "Autonomous Vehicles Lite Self-Driving Technologies Should Start Small, Go Slow," *IEEE Spectrum* 57, no. 3 (2020): 36-49.

24. Tang, J., Shaoshan, L., Pei, S., Zuckerman, S. et al., "Teaching Autonomous Driving Using a Modular and Integrated Approach," in *2018 IEEE 42nd Annual Computer Software and Applications Conference (COMPSAC)*, Vol. 1, Tokyo, Japan, 2018, 361-366, IEEE.

25. Liu, S., Gaudiot, J.L., and Kasahara, H., "Engineering Education in the Age of Autonomous Machines," *Computer* 54, no. 4 (2021): 66-69.

Chapter 08

Enabling Technologies—Cockpit Intelligence

The smart cockpit in electric vehicles isn't just an interface; it's a gateway where advanced technology meets driver comfort, transforming the driving experience into a harmonious blend of digital interaction and road awareness.

—Anonymous

The development of intelligent cockpit technologies for IEVs has seen substantial innovation and integration, particularly in enhancing HMIs and the overall in-vehicle experience. Central to these advancements are high-compute SoC platforms that facilitate more sophisticated interaction modes, improved human–computer interaction, and personalization features, as well as support for basic ADAS functions such as DMSs and surround view. In this chapter, we introduce the enabling technologies for cockpit intelligence.

8.1. Market Landscape

Cockpit intelligence technologies in IEVs focus on enhancing the interface between humans and machines, ensuring seamless integration of various functions such as navigation, infotainment, and driver assistance systems. Key players in the industry such as NVIDIA, Qualcomm, and Bosch are actively pushing the boundaries of what smart cockpits can do, incorporating advanced processors and AI capabilities to bring the automotive experience closer to consumer electronics in terms of connectivity and usability.

The NVIDIA DRIVE platform is designed to create a software-defined vehicle experience, enhancing the occupant experience alongside automated and autonomous driving features. It incorporates occupant understanding through interior-facing cameras, which monitor the state of the driver to ensure attention and safety. The system also personalizes the cockpit experience by adjusting settings such as seating, temperature, and infotainment according to the identified driver's preferences. Furthermore, it enhances security with features like biometric recognition, allowing vehicle access and personalization through driver identification. The DRIVE platform also supports advanced interaction capabilities, such as AR head-up displays (HUDs) and natural language processing, enabling drivers to receive vital driving information and control settings without distraction.

Qualcomm with their 8295 chip has enabled more complex functions such as 3-D HMI and car games, supporting higher-definition, smoother 3-D HMIs. Qualcomm has also supported the integration of intelligent cockpit functions with low-speed driving assistance. This platform leverages advanced AI compute and multicamera support capabilities to enable features such as 360° surround view and smart parking. Another key player, ECARX, utilizes AMD technology to enhance cockpit entertainment functions, supporting high-end graphics and immersive audio experiences, which allow for gaming and other high-demand applications directly from the vehicle's cockpit.

In the Chinese market, companies like Huawei have leveraged their expertise in telecommunications and software to develop intelligent cockpit systems using HarmonyOS, aiming to create a more interconnected and responsive in-car environment. This includes integration across different domains such as infotainment and body intelligent control, underlining the trend toward more integrated and versatile cockpit platforms.

Bosch has also been notable for its advancements in cockpit technologies, integrating multiple ECUs into a single cockpit domain controller to reduce vehicle cost and complexity, enhancing OTA capabilities. Their latest developments include intelligent cockpit solutions that integrate parking and driving functions, demonstrating the industry's move toward more centralized computing units that handle multiple aspects of vehicle operation.

These advancements underscore a significant shift toward intelligent cockpits that are not only functional but also pivotal in defining the driving experience, blending advanced computing, software innovation, and traditional automotive engineering to meet the evolving demands of consumers and the automotive industry.

8.2. Cockpit Intelligence Usage Scenarios

For the innovative usage scenarios of cockpit intelligence in IEVs, several products and systems currently implement these technologies, enhancing the driving experience and vehicle functionality.

- **Personalized Driver Settings:** Many modern vehicles, including those from Tesla, BMW, and Mercedes-Benz, feature advanced personalization capabilities. For instance, Tesla's Driver Profiles offer comprehensive personalization through driver profiles. Once a driver enters their vehicle, they can select their profile, and the vehicle automatically adjusts settings like mirror angles, seat positions, steering wheel height, and media preferences. This personalization extends to driving features such as acceleration and regenerative braking responses, tailoring the car's behavior to suit the driver's preferences.

- **Voice Control:** Voice control technologies deliver a few advantages. First, voice control minimizes the need for physical interactions with the vehicle's controls, allowing drivers to keep their eyes on the road and hands on the wheel. Second, by providing a hands-free method of interacting with the vehicle's systems, voice control makes driving more accessible to people with various physical abilities. Third, drivers can perform multiple functions without diverting attention from driving, from changing music tracks to adjusting the heating, ventilation, and air conditioning (HVAC) system, all through simple voice commands. For example, the Mercedes-Benz User Experience (MBUX) infotainment system features one of the most advanced voice control systems in the automotive industry. It includes a high-level natural language processing system that allows users to control the vehicle's functions by saying "Hey Mercedes," followed by a natural speech command. The system can handle complex queries and even understands indirect commands, significantly enhancing user interaction.
- **Gesture Control:** Similar to voice control, gesture control is also being integrated into more vehicles, providing a hands-free way to interact with vehicle systems without touching a screen or button. For example, BMW has been a pioneer in integrating gesture control within its vehicles. First introduced in models like the BMW 7 Series, the system allows users to perform a range of commands with simple hand gestures. For example, rotating a finger clockwise can increase the volume of the music system, while a swiping motion can accept or reject phone calls. This feature is part of BMW's broader iDrive system, which combines gesture control with touch and voice commands to provide a comprehensive user interface.
- **AR Displays:** AR display technologies provide a few benefits in IEVs. First, by keeping critical navigation and safety information within the driver's natural line of sight, AR displays reduce the need to look away from the road, decreasing driver distraction and increasing reaction times to road events. Second, AR technology transforms navigation by making it more intuitive. Directions and cues are displayed in the real-world environment, making them easier to understand and follow compared to traditional map-based navigation systems. Third, AR systems can dynamically adjust the information they display based on the context, such as changing driving conditions, traffic status, or the proximity of points of interest, providing a highly responsive and interactive user experience. For example, Audi has integrated AR into its HUD systems, particularly in models like the Audi Q4 e-tron. The AR HUD shows navigational cues and adaptive lane guidance, which appear as floating arrows or lines on the road, precisely where the driver needs to turn or change lanes. This system helps in reducing cognitive load and enhancing driver comprehension of navigation and safety-related information.

- **Health Monitoring:** The emerging usage scenario of health monitoring brings a few benefits: First, continuous monitoring of the driver's physical condition can help prevent accidents caused by health-related issues such as heart problems or fatigue. Immediate alerts can prompt the driver to take necessary actions, such as pulling over and resting, thereby reducing the risk of accidents. Second, by adjusting environmental variables in response to a driver's emotional state, systems like Mercedes-Benz's ENERGIZING Comfort contribute to a more comfortable and personalized driving experience. Third, regular monitoring can encourage healthier driving habits and highlight potential health issues that may need medical attention, contributing to better long-term health outcomes for drivers. Fourth, features that adjust vehicle settings to reduce stress can make commuting and long drives more pleasant and less taxing on the driver. One example is Mercedes-Benz ENERGIZING Comfort System, which uses sensors to adjust the vehicle's environment—including lighting, music, and seat massage functions—based on the detected stress or well-being of the driver. While not strictly a health monitoring system, it indirectly contributes to driver health by creating a more relaxing and stress-free driving environment.
- **Context-Aware Information and Assistance:** Cars like those from Tesla and BMW with advanced connectivity use location-based data and user habits to provide contextual assistance, such as suggesting navigation routes or nearby points of interest, or even adjusting vehicle settings for anticipated conditions. This feature brings a few benefits: First, automated adjustments to driving routes or vehicle settings based on the current context (such as weather, traffic, or time of day) streamline the driving experience, making it more convenient and enjoyable. Second, context-aware systems optimize routes and driving modes to improve fuel efficiency and reduce travel times, based on traffic patterns and road conditions. Third, these systems enhance comfort by learning and adapting to individual driver preferences, making each interaction more intuitive and responsive to the driver's needs. One example is Audi's AI-based Traffic Jam Pilot, which uses vehicle-to-infrastructure (V2I) communication to gather context about traffic conditions and infrastructure such as traffic light status or road signs. This system allows Audi vehicles to offer real-time driving assistance, like suggesting lane changes or speed adjustments in response to observed conditions, thereby enhancing safety and driving efficiency.
- **Seamless Integration with Smart Devices:** Modern EVs, such as those from Tesla and Rivian, offer extensive connectivity features that allow seamless integration with smartphones and other smart devices, enabling features like remote vehicle control, location tracking, and in-car Internet access. This brings a few benefits: First, drivers can interact with their vehicle as they do with their personal smart devices, using apps and services they are familiar with, reducing the learning curve and increasing usability. Second, integration technologies often come with advanced voice control features, allowing drivers to keep their hands on the wheel and eyes on the road while performing tasks like calling, texting, or setting navigation routes. Third, seamless integration allows the vehicle to access personal data from connected devices to tailor the driving experience to the user's preferences, such as playing preferred music or setting up the cabin environment according to the user's likes. Fourth, integration with

smart devices enables users to monitor their vehicle's status and perform certain functions remotely, enhancing security and convenience. One example is Apple CarPlay, which allows iPhone users to connect their devices to a car's infotainment system. This connection enables drivers to access key apps from their phones—such as navigation, messaging, music, and more—directly through the vehicle's interface.

These scenarios show how cockpit intelligence is becoming an integral part of the driving experience, providing not only enhanced comfort and convenience but also improved safety and efficiency. The integration of such advanced technologies is setting the stage for even more personalized and interactive vehicle environments.

8.3. Enabling Technologies

To enable the aforementioned scenarios in IEVs, the following common technologies, including HMI, digital instrument clusters, connectivity, and AI, are required [1, 2, 3, 4, 5]:

HMI: HMI plays a pivotal role in how users interact with their vehicles, blending technology and user experience to enhance safety, convenience, and enjoyment. HMI systems are sophisticated frameworks that manage the interaction between the driver, the vehicle, and its systems, incorporating touch interfaces, voice commands, gesture control, and more. Key components of HMI in IEVs include the following.

Touchscreens are the central element of most modern IEV HMIs. They provide access to navigation, climate control, entertainment systems, vehicle settings, and more. These screens often feature customizable interfaces that can be tailored to show information relevant to the driver's preferences and needs. Voice control allows drivers to perform a multitude of tasks without taking their hands off the wheel. These systems are designed to learn from user inputs to improve accuracy and responsiveness over time. Gesture control is being integrated into more vehicles to provide an even more seamless interaction experience. The gesture control system allows drivers to command certain functions with simple hand movements, reducing the need to physically touch controls or screens, which helps minimize driver distraction. AR HUDs project important information, such as speed, navigation arrows, and safety warnings, directly onto the windshield in the driver's line of sight. This technology helps keep drivers' eyes on the road, significantly enhancing safety.

HMI design must balance complexity and usability to avoid overwhelming or distracting the driver. Safety is a paramount concern, as poorly designed HMIs can lead to increased cognitive load and distraction. Additionally, as vehicles become more connected, security concerns, especially related to data privacy and system integrity, become increasingly important.

Digital Instrument Clusters: Digital instrument clusters represent a significant evolution in how drivers interact with their vehicles, offering enhanced functionality, customization, and a futuristic aesthetic compared to traditional analog clusters. These digital displays are pivotal in delivering a wide array of information to the driver in a clear, concise, and visually appealing manner. Key features of digital instrument clusters include the following.

First, digital clusters utilize high-resolution, full-color displays that can show a wide range of dynamic information beyond just speed and fuel gauge. This includes navigation maps, battery charge status, energy consumption metrics, and even infotainment system data. The use of OLED and LCD technology allows these displays to offer sharp visuals and customizable layouts.

Second, one of the standout features of digital instrument clusters is their ability to be customized according to the driver's preferences. For instance, drivers can choose what information is displayed, where it is located on the dashboard, and how it is presented. This level of personalization enhances the driving experience by tailoring the interaction to fit the individual's needs.

Third, digital clusters are often integrated with the vehicle's central infotainment system and other ADASs, providing seamless interaction across devices. For example, if the navigation system detects an upcoming turn, the instrument cluster can highlight this information prominently to assist the driver without needing to look away to a different screen.

Fourth, these clusters can enhance safety by incorporating alerts directly into the driver's line of sight. This includes lane departure warnings, speed limit indicators, and proximity alerts that integrate with the vehicle's sensors and cameras. By centralizing these data in front of the driver, it reduces the need to look away from the road, thus minimizing distractions.

Telematics and Connectivity: Telematics and connectivity enhance vehicle functionality, safety, and user experience through advanced communication and data exchange. These systems integrate telecommunications and informatics to monitor and manage vehicle and driver data, facilitating a wide range of services from remote diagnostics to real-time traffic updates. Core technologies in telematics and connectivity include the following:

First, telematics systems are built directly into vehicles and typically include a dedicated SIM card and modem to enable connectivity. These systems can independently connect to the Internet, allowing for features such as emergency calling, stolen vehicle tracking, and remote control of vehicle functions like preconditioning the climate or unlocking doors.

Second, many telematics systems are designed to work seamlessly with smartphones, extending the vehicle's connectivity features. Technologies like Apple CarPlay, Android Auto, and proprietary apps allow drivers to connect their devices to the vehicle's system, enabling them to use their phone's interface, apps, and data plan for streaming music, navigation, and more.

Third, vehicle-to-everything (V2X) communication is an advanced form of telematics that enables vehicles to communicate with each other and with road infrastructure. This technology improves road safety by allowing vehicles to exchange information about their speed, direction, and road conditions, potentially preventing accidents by alerting drivers to hazards their sensors might not detect.

Fourth, modern telematics systems often utilize cloud-based platforms to analyze large volumes of data collected from the vehicle. These data can provide insights into vehicle performance, maintenance needs, and driving patterns. Cloud connectivity also supports OTA updates, which allow manufacturers to update software without requiring a visit to a dealership.

AI and Machine Learning (ML): AI and ML are increasingly being integrated into IEVs to provide various features and services.

These technologies enable vehicles to learn from vast amounts of data, make intelligent decisions, and improve over time through adaptive algorithms. Key applications of AI and ML in IEVs include the following:

First, ML models are used to predict vehicle maintenance needs by analyzing real-time data from various sensors in the vehicle. These models can detect anomalies and predict failures before they occur, thereby reducing downtime and maintenance costs. For example, companies like BMW and Tesla use predictive analytics to provide vehicle health reports and maintenance alerts to drivers, improving the reliability and longevity of the vehicle components.

Second, AI in IEVs extends to personalizing the driving experience. Through learning algorithms, vehicles can adapt to the preferences and habits of their drivers, automatically adjusting settings such as seat position, temperature, and even driving styles. Systems can also learn to prioritize information based on what is most frequently used or needed by the driver.

Third, ML algorithms optimize energy consumption and battery usage in EVs. These algorithms process data from past usage patterns and real-time conditions to efficiently manage battery life, regenerative braking, and charging strategies to maximize range and battery health.

8.4. Integration of LLMs into IEVs

We believe the future of IEV intelligence lies in LLMs. LLMs are going to enable a broad spectrum of functionalities that enhance both the driving experience and vehicle intelligence. Their profound ability to process and generate human-like text, understand natural language and multimedia data, and learn from vast amounts of sensor data allows LLMs to be integral in several key automotive technologies. In this section, we provide an introduction to LLMs as well as review potential applications of LLMs in IEVs.

8.4.1. Development of LLM

The development of LLMs has been a pivotal aspect of advancements in AI, particularly in the field of natural language processing [6, 7]. The journey began with simpler models and has progressed to sophisticated systems capable of understanding and generating human-like text. Early models were focused on basic tasks like spelling correction and keyword search but lacked a deeper understanding of language semantics.

The breakthrough came with the introduction of ML techniques that use large amounts of data to train models. This shift started with statistical methods and moved toward neural network-based models in the early 2010s, which significantly improved the ability of machines to process and understand complex language patterns. The real transformation in LLMs began with the development of models like Google's Bidirectional Encoder Representations from Transformers (BERT) and OpenAI's Generative Pre-trained Transformer (GPT) series [8, 9].

BERT revolutionized the understanding of the context in the text by processing words in relation to all the other words in a sentence, rather than one-at-a-time in order. This allowed for a much richer understanding of language context and nuance. OpenAI's GPT series further advanced this field by introducing a few-shot learning ability, where the model can adapt to new tasks with minimal examples. GPT-3, the third iteration in the series, demonstrated the capacity of these models to generate coherent and contextually relevant text across a wide range of subjects and formats, from conversational dialogue to summarizing complex documents.

Currently, LLMs are capable of performing a variety of language-based tasks that were previously thought to be exclusive to humans. These include writing articles, composing poetry, coding software, and even generating creative fiction. The models achieve this by processing vast datasets of text from the Internet to learn language patterns, grammar, facts about the world, and even style and tone.

The future outlook for LLMs suggests even greater integration into everyday technology. Potential developments include more refined and specialized models tailored to specific professional fields such as law, medicine, and academia, where they can assist with research and documentation. Another anticipated advancement is the improvement in multilingual and culturally aware models, which can understand and generate text in multiple languages with cultural nuance.

However, this rapid growth also brings challenges, particularly in areas like privacy, security, and ethical considerations regarding the use of generated text and the potential for misuse. As these models continue to evolve, they are expected to become more integral to our digital interactions, influencing how we interact with machines and manage information in our daily lives. The development trajectory of LLMs showcases a promising yet cautious path forward in the realm of AI.

8.4.2. LLMs in IEVs

Deploying LLMs in IEVs enables many new features, enhancing both the functionality and the user experience:

- **Autonomous Driving:** LLMs improve the autonomous driving systems by enabling more complex decision-making and a better understanding of the driving context. By processing natural language inputs, LLMs can interpret instructions and environmental data, allowing vehicles to react more intelligently in real-time situations [10, 11]. The integration of LLMs with multimodal data inputs such as images, videos, and other sensor data greatly enhances the perception systems of autonomous vehicles. Unlike previous deep learning models, LLMs are able to process and respond to complex multimodal data, thereby improving the safety and efficiency of autonomous vehicles. However, several technical challenges remain, particularly the integration of LLMs with real-time vehicle systems, data privacy, and the need for robust multimodal datasets. Future research directions include enhancing the capabilities of LLMs to understand and interact with the environment more naturally and effectively. In addition to on-vehicle intelligence, the applications of LLMs in traffic systems can assist in navigation, real-time decision-making, and user interaction by processing extensive data from vehicle sensors and external environments.
- **Advanced Interaction and Infotainment Systems:** LLMs facilitate more natural interactions between drivers and their vehicles, significantly improving the user interface. Vehicles become more responsive to voice commands and can provide real-time feedback, advice, and even entertainment options,

creating a more engaging and personalized driving experience [12, 13]. Particularly, LLMs have been utilized to empower in-car conversations. The study [12] addresses current limitations in proactive interaction and user intent recognition, proposing a novel framework and strategy to improve these systems' effectiveness and user satisfaction. The paper emphasizes the importance of proactive interactions to reduce distractions and enhance driving safety, aligning with users' cognitive needs. The paper also develops a strategy to empower LLMs to handle user interactions more effectively, which involves rewriting user input to better fit driving contexts, reacting by generating proactive responses, and reflecting to ensure the responses align with the desired proactivity level. Experimental results confirm that the LLM-based system demonstrated a high success rate in understanding and responding to user commands, outperforming traditional models. It also showed promising results in real-world tests, significantly reducing the need for human intervention.

- **Sophisticated Navigation Systems:** Navigation and route planning are significantly improved with the deployment of LLMs. These models can process complex instructions and integrate real-time environmental data to optimize routes, reduce travel times, and avoid hazards, enhancing both efficiency and safety [14, 15]. Particularly, [15] presents a comprehensive study on the use of LLMs for solving vehicle routing problems (VRPs) using natural language descriptions. The research explores how LLMs, particularly GPT-4, can be applied to generate executable code from natural language task descriptions to solve VRPs. This is significant as it bypasses the traditional need for expert human intervention to translate these problems into computational models. The study benchmarks the performance of LLMs like GPT-4 against VRPs using metrics of feasibility, optimality, and efficiency. GPT-4 showed promising results, indicating the practical viability of using LLMs in this domain. This research underscores the potential of LLMs to transform how complex logistical problems are approached and solved, leveraging the power of advanced AI to enhance efficiency and reduce reliance on human experts in the field of vehicle routing.
- **Continuous Learning and Adaptation:** Vehicles equipped with LLMs can continuously learn and adapt to a driver's preferences and driving habits, enabling personalized adjustments to vehicle settings, driving modes, and infotainment preferences. This ongoing adaptation is facilitated by the AI's ability to learn from the data generated during drives, which helps in enhancing the driving experience over time. The study [16] explores the application of LLMs to improve the personalization of robotic tasks. The research focuses on enabling robots to learn and generalize individual user preferences for object placement from minimal examples. The study demonstrates how LLMs can be employed to teach robots personalized preferences for where items should be placed, based on a few input examples provided by users. This involves using LLMs to summarize these examples into general rules that robots can apply in future tasks. The robot uses few-shot learning to understand and generalize user preferences with high accuracy. This approach allows the robot to perform tasks according to personalized preferences with minimal initial input. This research indicates a significant advancement in using AI to enhance personalized robotics, making them more adaptable and useful for everyday tasks in dynamic human environments.

8.5. Summary

The chapter delves into the sophisticated developments and integrations of cockpit technologies. It emphasizes how these advancements transform the driving experience by enhancing HMIs. These technologies facilitate a more intuitive interaction between the driver and the vehicle, supporting not only entertainment and comfort but also crucial driving functions. In essence, the chapter outlines how the convergence of advanced computing, software innovation, and traditional automotive engineering is crafting more responsive, integrated, and user-centric cockpit environments in IEVs. These enhancements not only improve the user experience but also align with broader trends in automotive technology, emphasizing safety, efficiency, and personalized user interactions. Particularly, we would like to emphasize that the future of cockpit intelligence lies in the development of LLMs and the integration of these models into IEV platforms.

References

1. Kun, A.L., "Human-Machine Interaction for Vehicles: Review and Outlook," *Foundations and Trends® in Human–Computer Interaction* 11, no. 4 (2018): 201-293.
2. Meixner, G., Häcker, C., Decker, B., Gerlach, S. et al., "Retrospective and Future Automotive Infotainment Systems—100 Years of User Interface Evolution," in: Meixner, G. and Müller, C. (eds), *Automotive User Interfaces: Creating Interactive Experiences in the Car* (Cham: Springer, 2017), 3-53.
3. Rameau, F., Ha, H., Joo, K., Choi, J. et al., "A Real-Time Augmented Reality System to See-Through Cars," *IEEE Transactions on Visualization and Computer Graphics* 22, no. 11 (2016): 2395-2404.
4. Lu, N., Cheng, N., Zhang, N., Shen, X. et al., "Connected Vehicles: Solutions and Challenges," *IEEE Internet of Things Journal* 1, no. 4 (2014): 289-299.
5. Hylving, L. and Schultze, U., "Evolving the Modular Layered Architecture in Digital Innovation: The Case of the Car's Instrument Cluster," in *International Conference on Information Systems*, Milan, Italy, 2013.
6. Chang, Y., Wang, X., Wang, J., Wu, Y. et al., "A Survey on Evaluation of Large Language Models," *ACM Transactions on Intelligent Systems and Technology* 15, no. 3 (2024): 1-45.
7. Wei, J., Tay, Y., Bommasani, R., Raffel, C. et al., "Emergent Abilities of Large Language Models," arXiv preprint arXiv:2206.07682, 2022.
8. Devlin, J., Chang, M.W., Lee, K. and Toutanova, K., "Bert: Pre-Training of Deep Bidirectional Transformers for Language Understanding," arXiv preprint arXiv:1810.04805, 2018.
9. Brown, T., Mann, B., Ryder, N., Subbiah, M. et al., "Language Models Are Few-Shot Learners," *Advances in Neural Information Processing Systems* 33 (2020): 1877-1901.
10. Cui, C., Ma, Y., Cao, X., Ye, W. et al., "A Survey on Multimodal Large Language Models for Autonomous Driving," in *Proceedings of the IEEE/CVF Winter Conference on Applications of Computer Vision*, Waikoloa, HI, 2024, 958-979.
11. Fu, D., Li, X., Wen, L., Dou, M. et al., "Drive Like a Human: Rethinking Autonomous Driving with Large Language Models," in *Proceedings of the IEEE/CVF Winter Conference on Applications of Computer Vision*, Waikoloa, HI, 2024, 910-919.
12. Du, H., Feng, X., Ma, J., Wang, M. et al., "Towards Proactive Interactions for In-Vehicle Conversational Assistants Utilizing Large Language Models," arXiv preprint arXiv:2403.09135, 2024.
13. Ondas, S. and Gurcik, M., "Domain-Specific Language Models Training Methodology for the In-Car Infotainment," *Intelligent Decision Technologies* 11, no. 4 (2017): 417-422.
14. Liu, Y., Wu, F., Liu, Z., Wang, K. et al., "Can Language Models Be Used for Real-World Urban-Delivery Route Optimization?" *The Innovation* 4, no. 6 (2023): 100520.
15. Huang, Z., Shi, G. and Sukhatme, G.S., "From Words to Routes: Applying Large Language Models to Vehicle Routing," arXiv preprint arXiv:2403.10795, 2024.
16. Wu, J., Antonova, R., Kan, A., Lepert, M. et al., "Tidybot: Personalized Robot Assistance with Large Language Models," *Autonomous Robots* 47, no. 8 (2023): 1087-1102.

Chapter 09

Enabling Technologies: Ecosystem

Technology alone is not enough. It's technology married with liberal arts, married with the humanities, that yields the results that make our hearts sing.

—Steve Jobs

We start this chapter by embarking on a historical journey to understand the pivotal role of ecosystems in technological evolution, using the smartphone industry as an example, as we believe that the IEV industry is currently repeating the journey of evolution of the smartphone industry. As illustrated in Figure 9.1, the early 2000s were dominated by feature phones, devices with limited functionality that devoted approximately 90% of their computing power to basic communication tasks, such as encoding and decoding. This left a mere 10% of computing power for applications, severely restricting the diversity of applications and limiting the mobile computing ecosystem's market size to around $10 billion [1].

The emergence of smartphones represented a significant paradigm shift, sparking an unprecedented demand for more computing power to accommodate a growing multitude of mobile applications. This demand led to the evolution from single embedded chips to complex mobile SoC, featuring multicore CPUs, mobile GPUs, mobile DSPs, and sophisticated power management systems. As a result, up to 90% of a smartphone's computing power became dedicated to applications, enabling the rise of platforms like YouTube, WhatsApp, Uber, and others. This technological advancement expanded the mobile computing ecosystem's market size to an impressive $800 billion today.

Figure 9.1 The growth of mobile computing ecosystem.

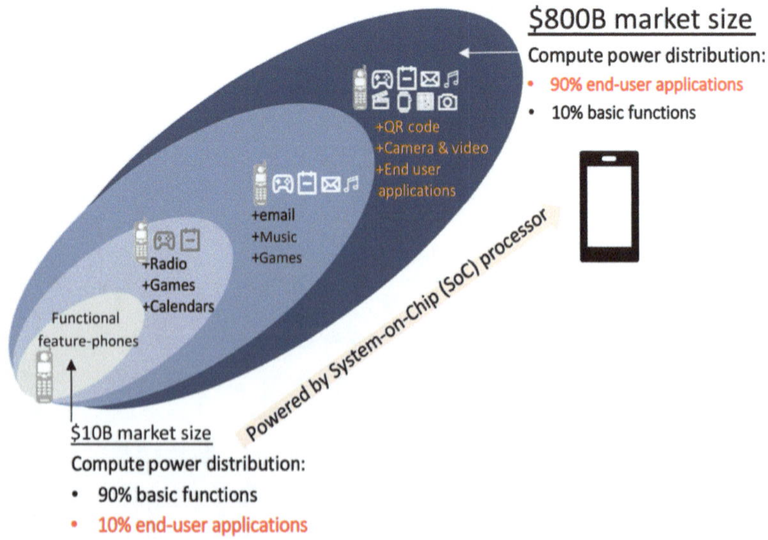

This historical trajectory highlights the profound impact of ecosystems on technological progress. Currently, we stand at the beginning of the IEV era, with the IEV ecosystem still in a rapid phase of evolution. We anticipate the emergence of many innovative applications on IEV platforms. This chapter focuses on the application of AMCs [2, 3] as a case study to elucidate the technologies that enable medical applications on IEVs, showcasing the continued transformative potential of ecosystems in shaping our technological landscape.

9.1. Usage Scenario of In-Vehicle Healthcare Services

In this section, we use acute myocardial infarction and stroke as examples to explain why in-vehicle healthcare services (IVHSs) are a more efficient way of healthcare service delivery. Emergency situation is one healthcare use case in vehicles. For instance, accessing immediate medical help when a driver is having a heart attack is imperative to saving his or her life. The long waiting time and transition time in the ED affect the health outcome of the patient profoundly, often resulting in loss of lives. Also, the ED is strongly connected with other departments for hospital admissions and transition. Waiting time and transition time represent the amount of accessing hospital services and are some of the most important performance indicators. The current outpatient service system faces challenges regarding the effective arrangement of appointments and providing medical services to patients. Multiple researches have confirmed that patients have experienced serious health complications due to prolonged ED waiting times.

On the hospital side, according to the US National Survey from 2015, more than 90% of US hospitals claim to be over capacity, and over 50% of the hospitals also experienced delays in transition times from patients' homes to the ED.

In May 2014, the Centers for Disease Control and Prevention (CDC) reported average ED waiting times (approximately 30 min) and home–hospital transition times (approximately 120 min), which add up to roughly 2 h in the ER scenarios, resulting in loss of lives. The long transition times and heavy emergency resource demand have led to a number of problems, including ambulance refusals, prolonged patient waiting times, increased suffering for those who wait, rushed and unpleasant treatment environments, and potentially poor patient outcomes.

IVHS is the enabling technology for anywhere anytime healthcare access, as IVHS not only continuously monitors the driver's or rider's health status, but immediately connects the driver/rider to medical experts should an emergency occur. The first benefit of IVHS is that it effectively eliminates the waiting time, and through telemedicine, many ED triage functions can be performed in-vehicle. In addition, by having autonomous driving capability, more than 30% transportation time reduction has been demonstrated, and thus autonomous vehicles with IVHS capability significantly reduce the transition time, further improving the efficiency of healthcare service delivery.

To understand the impact of IVHS, we utilize the cost-effectiveness analysis methodology, a common practice in the medical field to simulate 100 patients who are located within 20 km of a major hospital with ED settings. We choose mean time (in minutes) from patient-reported symptom onset time to time of entering the ED as our primary indicator, with the secondary process indicators of process velocity and process waste.

For the cost of myocardial infarction and stroke in ED, we use the mean data provided by the US CDC, which is $30,000 USD. For the IVHS, we assume the worst-case scenarios, doubling the cost with usual care patients, resulting in a total cost of $60,000 USD for this group. The results from this analysis study indicate that IVHS significantly outperforms regular healthcare delivery methods in terms of efficiency, mainly due to the elimination of waiting time and significant reduction of transition time.

9.2. Automotive Health Domain Controllers

Now that we have reviewed the special usage scenarios of IEVs in healthcare, what technologies are demanded to enable these scenarios? In automotive applications, a domain controller is a computer that controls a set of vehicular functions related to a specific area or domain. Functional domains requiring domain controllers are typically compute-intensive and connect to many input/output (I/O) devices. Examples of relevant domains include autonomous driving, smart cockpit, and chassis. For instance, an autonomous driving domain controller receives inputs from sensors around the vehicle, such as radars and cameras, and uses that input to create a model of the surrounding environment. Autonomous driving software in the domain controller then makes planning decisions to generate actions the vehicle should take.

In this section, we discuss the necessity for automotive health domain controllers (AHDCs) to empower IVHS and define the core features of an AHDC. Intelligent vehicles are the best space to deliver healthcare services for three reasons: First, in many countries, people spend a considerable amount of time in cars. The average daily time is about 1 h in the United States. This allows continuous monitoring of one's health conditions. Second, intelligent vehicles are one of the best-equipped environments in our daily lives for health monitoring. Third, vehicles are private spaces suitable for healthcare service delivery. Yet, intelligent vehicles have not been utilized for healthcare service scenarios due to the lack of an AHDC to integrate all necessary technologies to enable a seamless healthcare service experience for the drivers and riders.

Particularly, AHDCs will enable **built-in** healthcare features by integrating healthcare sensors in the vehicle manufacturing process. AHDCs will support **brought-in** healthcare features by integrating with healthcare devices that the drivers and passengers carry with them into vehicles, such as smartphones and smartwatches. AHDCs will facilitate **beamed-in** or cloud-enabled healthcare services through reliable connectivity [4].

An AHDC consists of a healthcare **I/O interface**, along with a data specification, to integrate both the built-in and the brought-in healthcare sensors. With the influx of heterogeneous healthcare sensing data, a health AI engine performs automatic abnormality detection, prognosis, and sensor data analysis. The **health AI engine** is key to enabling future in-vehicle AI doctors to improve healthcare efficiency. Then, a **storage engine** along with encryption capability is required to store the private and sensitive healthcare data, both raw sensing data and AI prognosis data. The storage engine is to guarantee that the patient owns his healthcare data in case he chooses not to upload the data to the cloud. Next, a c**onnectivity engine** is required to enable beamed-in or cloud-based healthcare services, such as telemedicine and the transmission of health sensing data. The key challenge for AHDC is to integrate all these technical pillars into a domain controller with limited compute power, energy consumption, and cost, but with the strictest reliability requirements.

In the past few years, the industry has explored integrating healthcare usage scenarios into the driving and riding experiences. For instance, Ford has developed an electrocardiography (ECG) reader that is integrated into the driver's seat. Toyota has developed a future concept vehicle to demonstrate the mobility service of allowing a passenger to talk with a doctor while en route and undergo a medical exam on their way to the hospital. This vehicle goes where it is needed, including a customer's home or other destinations.

One challenge is to integrate various healthcare sensors into intelligent vehicles to deliver a seamless healthcare delivery experience. In particular, the sensor and equipment deployment needs to be highly customized to the specific healthcare service scenarios. Taking acute myocardial infarction and stroke as examples, this deployment involves built-in sensors, including millimeter-wave radars, IR cameras, thermal cameras, photoelectric sensors, conductive textiles, as well as brought-in sensors, such as mobile phones and wearable devices. In the scenario of acute myocardial infarction, vital sign monitoring tracks a patient's day-to-day vital information, generates basic health reports, and identifies any abnormality.

Facial expression tracking specifically targets symptoms such as facial palsy, a sign of stroke. When signs of facial palsy are detected, the vehicle will issue a warning and may connect to a remote doctor for symptom confirmation. An infarction warning goes one step further to confirm the patient's medical situation. When a patient gets to the infarction warning stage, he or she may already be unconscious. The intelligent vehicle with an autonomous driving function may directly send the patient to a nearby ED while sending the healthcare data along the way, or the intelligent vehicle summons an ambulance to save the patient's life as soon as possible.

If one scenario involves so many advanced sensors, we can foresee that an extensive number of sensors are required to provide a comprehensive healthcare service, especially with the rapid development of in-vehicle healthcare sensor technologies. A standard method of integrating and making sense of the enormous amount of sensing data is required. Health Level 7 (HL7) refers to a set of international standards for the transfer of healthcare data between software applications used by various healthcare providers.

These standards focus on the application layer, which corresponds to the seventh layer in the OSI model. The HL7 standards have been adopted by other standards-issuing bodies such as the American National Standards Institute and the International Organization for Standardization. By having all sensor data conform to the HL7 standards, we can enable seamless integration of various sensors into intelligent vehicles and facilitate the interoperability of healthcare data.

9.3. Edge AI for Healthcare Services

There are multiple benefits of performing edge AI for healthcare services in vehicles. First, many users do not want to have their medical data uploaded to the cloud due to privacy concerns. By integrating edge AI in intelligent vehicles, the raw medical data of the patient can be consumed locally, thus effectively addressing the privacy concerns. Second, edge AI is connectivity-prone. In many locations with limited network connectivity, edge AI is still able to provide healthcare services, thus guaranteeing quality of service under different situations. Third, and the most important reason, edge AI is able to greatly improve healthcare service efficiency, as edge AI can perform prognosis for the doctors and thus significantly increase doctors' service bandwidth.

Due to the recent rapid development of deep learning technologies, many deep learning-enabled medical applications have emerged [5, 6]. To name a few, skeleton-based action recognition techniques can be utilized to detect whether a driver or a rider in a vehicle is suffering from an emergency, such as stroke. Real-time electrocardiogram (ECG) analysis can be utilized for anxiety attack detection. In addition, deep learning techniques can be applied to multimodal data for early detections of Parkinson's disease.

As most medical AI algorithms are compute-intensive and energy-hungry, one challenge is to integrate a diverse set of AI algorithms onto compute- and energy-constrained automotive-grade compute devices. To address this exact problem, a software technique dubbed compression and compilation co-optimization can be utilized to enable AI applications on embedded devices [7, 8]. Specifically, the first step, model compression, aims to reduce the size and improve the speed of deep learning models. Compression techniques fall into two categories: pruning and quantization. Pruning removes layers or convolution filters or channels, while quantization reduces the precision of parameters (e.g., from floating-point to short integer). The second step, compilation, refers to the process of generating executable codes from a given deep learning model, effectively mapping the high-level operations in deep learning to the low-level instructions that the underlying hardware supports. Compilation plays a critical role in optimizing the code for efficient executions. This technique has been demonstrated to accelerate deep learning models by up to 180× and has enabled many compute-intensive applications. Compression and compilation co-optimization and similar techniques are the core of AHDC's fundamental systems software layer so as to integrate various edge AI algorithms for healthcare services.

9.4. Connectivity Requirements

Connectivity is key to guarantee the QoS of in-vehicle applications, especially for healthcare services. There are two categories of data communication for IVHSs: healthcare sensing data and telemedicine data stream. Healthcare sensing data involve the transmission of raw sensing data collected from healthcare sensors, such as imaging devices. This type of communication is bandwidth-constrained but less latency-sensitive. It demands a communication bandwidth of 15 MB/s or more and a latency of 1 sec or less. Telemedicine data stream is mostly HD video conferencing workloads to allow interactive communication between the patient and the remote doctor. This type of communication is highly latency-sensitive; it demands a communication bandwidth of 3 MB/s or more and a latency of 100 ms or less.

However, through our measurements in the field, the real-world communication conditions are not able to satisfy these requirements. We have tested three commercial networks in China (a country with great communication infrastructure) with a moving vehicle: 4G, 5G, and LTE V2X. Note that LTE V2X is used for communicating short messages, whereas 4G and 5G mobile networks are used for transferring a large volume of video streaming and healthcare sensing data. While 3GPP (3rd Generation Partnership Project) has recently completed 5G-NR release 16 to enhance vehicle-to-everything (V2X) direct communication, the state-of-the-art commodity onboard unit (OBU) and roadside unit (RSU) only support LTE V2X direct communication (specified by 3GPP release 14), of which communication capacity is insufficient for in-vehicle healthcare scenarios.

The results show that the 4G network enjoys the best coverage, a 94% deployment rate in China, but the 4G network is not able to satisfy the bandwidth and latency requirements. The 5G network, although it satisfies the latency and bandwidth requirements in the average case, has much less coverage. The V2X network delivers extremely low latency but fails to deliver high bandwidth, and it is only deployed in very limited areas. To exacerbate the communication problem, we have observed very serious latency and bandwidth fluctuation problems in the 4G and the 5G networks, and thus the current commercial communication networks are not able to guarantee QoS of IVHSs.

Although no existing commercial network is able to guarantee the QoS of IVHSs, each network has its own advantage, whether it is coverage, latency, or bandwidth. To address the QoS problem, a mechanism is required to get the best of each network dynamically. We have developed a multimodal communication middleware (MMCM) to dynamically dispatch communication requests to different communication channels (4G, 5G, or V2X) [9]. MMCM constantly monitors the conditions (latency, bandwidth) of each network, and then as a communication request comes in, MMCM dispatches the request to the most suitable network. Note that each communication request comes with attributes, including priority, latency sensitivity, and size. Then, policies have been implemented to allocate network resources.

Initial results have demonstrated that MMCM has outperformed existing middleware, such as ROS 2, or raw TCP/IP communication that relies on one network, by 5× in terms of communication latency. This has greatly improved QoS of IVHSs. Based on this result, we believe that MMCM is a key enabling technology dealing with the wild real-world network communication problems, and MMCM should be standardized to support various in-vehicle healthcare applications.

9.5. Need for Standardization

In the previous sections, we have delved into the specific technology areas to enable IVHSs through AHDCs. Standardization in AHDCs is the imminent and imperative next step as the industry marches toward the grand vision of a mobile healthcare system, especially in the following areas:

- **Healthcare hardware qualification:** Automotive qualification means that a component or system meets specific industry standards for manufacturing and performance. For example, in AEC-Q100, for grade 2 semiconductor components, high-temperature operating life testing of 1000 h at a temperature of 125°C is required. However, most of the healthcare sensors and devices are not designed according to automotive grade requirements. Hence, the automotive community and the healthcare device community should join forces to develop standards for on-vehicle healthcare devices.
- **Healthcare data communication in vehicle:** HL7 provides a broad set of standards creating domain-specific protocols to achieve interoperability among healthcare service providers. However, HL7 only focuses on high-level data interoperability, without defining the full stack of health data exchange in the vehicular communication network. Particularly, encryption is an essential feature for healthcare data exchange, but there is still no standardization on how vehicular communication network can provide built-in support for healthcare data communication needs.
- **Healthcare AI software:** AI technologies have the potential to transform healthcare by deriving new and important insights from the vast amount of data generated during the delivery of healthcare every day. The regulatory efforts on healthcare AI software are still in their infancy, let alone the scenarios of delivering healthcare services with AI in vehicles. As an immediate next step, a cross-industry joint effort is required for standardization of AI-enabled IVHS delivery.
- **Network communication:** Efficient and reliable network communication is essential in guaranteeing the quality of healthcare service in vehicles. However, the existing real-world communication networks are unsatisfactory. As healthcare service communication is mission-critical, there should be standardization in commercial communication networks to provide high priority for healthcare service connections. Also, there should be standardization on the multimodal communication backbone to guarantee QoS.

9.6. Summary

In this chapter, we have drawn parallels between the emergence of IEVs and the evolution of the smartphone industry. Just as smartphones revolutionized mobile computing, IEVs are poised to redefine the automotive landscape, offering a myriad of innovative applications across various domains. While our focus has been on the application of IVHSs, particularly through the lens of AMCs, it's essential to recognize the broader spectrum of potential usage scenarios for IEVs. From AHDCs to edge AI for healthcare services, the technological advancements we've explored underscore the vast potential for IEVs to become multifaceted platforms for delivering a wide range of services beyond transportation.

As we envision the future of IEVs, it becomes increasingly clear that their impact will extend far beyond the realm of mobility, ushering in a new era of interconnectedness and accessibility. From enabling seamless healthcare delivery to fostering innovative solutions for various industries, IEVs represent a convergence of technology, connectivity, and human ingenuity. As we navigate this technological frontier, collaboration and standardization will be paramount in realizing the full potential of IEVs across diverse usage scenarios.

References

1. Liu, S., "Shaping the Outlook for the Autonomy Economy," *Communications of the ACM* 67, no. 6 (2024): 10-12.
2. Liu, S., Kong, A., Huang, Y., and Liu, X., "Autonomous Mobile Clinics," *Bulletin of the World Health Organization* 100, no. 9 (2022): 527.
3. Liu, S., Huang, Y., and Shi, L., "Autonomous Mobile Clinics: Empowering Affordable Anywhere, Anytime Healthcare Access," *IEEE Engineering Management Review* 50, no. 4 (2022): 147-154.
4. Liu, S., Huang, Y., Kong, A., Tang, J. et al., "Rise of the Automotive Health-Domain Controllers: Empowering Healthcare Services in Intelligent Vehicles," *IEEE Internet of Things Journal* 9, no. 24 (2022): 24882-24889.
5. Yang, T., Yang, T., Liu, A., An, N. et al., "AICOM-MP: An AI-Based Monkeypox Detector for Resource-Constrained Environments," *Connection Science* 36, no. 1 (2024): 2306962.
6. Yang, T.T., Yang, T.T., An, N., Kong, A. et al., "AI Clinics on Mobile (AICOM): Universal AI Doctors for the Underserved and Hard-to-Reach," arXiv preprint arXiv:2306.10324, 2023.
7. Guan, H., Liu, S., Ma, X., Niu, W. et al., "CoCoPIE: Enabling Real-Time AI on Off-the-Shelf Mobile Devices via Compression-Compilation Co-Design," *Communications of the ACM* 64, no. 6 (2021): 62-68.
8. Tang, J., Sun, D., Liu, S., and Gaudiot, J.L., "Enabling Deep Learning on IoT Devices," *Computer* 50, no. 10 (2017): 92-96.
9. Wu, T., Liu, S., Yu, B., Wang, S. et al., "INTERNEURON: A Middleware with Multi-Network Communication Reliability for Infrastructure Vehicle Cooperative Autonomous Driving," arXiv preprint arXiv:2210.15939, 2022.

Chapter 10

Semiconductors for Intelligent Electric Vehicles

The semiconductor is the brain of industry. Any product that does not contain a semiconductor is not a product of the 21st century.

—Anonymous

Similar to its predecessors personal computing and mobile computing, autonomous machine computing (AMC) is the core technology stack that empowers a wide range of autonomous machine form factors, including Mars rovers, intelligent vehicles, autonomous drones, delivery robots, home service robots, agriculture robots, industry robots, and many more that we have yet to imagine [1, 2]. AMC involves sensing technologies, computing technologies, communication technologies, autonomous machine algorithms, reliability and security, and many other technical areas. As of today, AMC is still evolving and being defined. In this chapter, we review semiconductors for IEVs, especially the research results of AMC that will enable future generations of IEVs.

10.1. Historical Insights—The Virtuous Cycle

To elucidate the impact of AMC on the evolution of the autonomy economy, let us examine historical precedents. Figure 10.1 illustrates the correlation between the market sizes of ecosystems and their corresponding semiconductor markets in the realms of personal and mobile computing. Currently, the mobile processor market is valued at $35.1 billion, with the mobile computing ecosystem's market size soaring to $800 billion—approximately 23 times larger. Similarly, the personal computing processor market stands at $55 billion, with its ecosystem valued at $900 billion, making it 16 times the processor market's size [3]. This pattern indicates that, as the computing era matures, the semiconductor industry plays a crucial role, potentially fostering an ecosystem that is 15 to 25 times its own market size.

Figure 10.1 The virtuous cycle of computing technologies and their ecosystems.

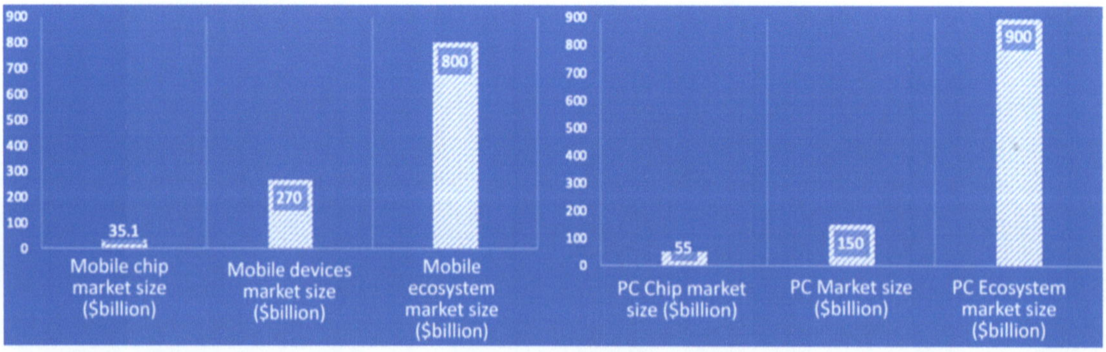

Courtesy of Shaoshan Liu.

These historical insights reveal a fundamental truth: the semiconductor industry acts as a cornerstone for modern economies. The growth of nascent sectors like AMC prompts the semiconductor industry to innovate and develop necessary technologies. These advancements, in turn, fuel further growth in the emerging sector, creating a self-reinforcing cycle of development and expansion. This virtuous cycle underscores the semiconductor industry's pivotal role in enabling and propelling the growth of sectors critical to the future economy, such as AMC in the burgeoning autonomy economy.

Under the central thesis that the semiconductor industry is the key enabler of modern economy, we believe that the growth of an emerging sector, such as AMC, will incentivize the semiconductor industry to develop enabling technologies; and the newly developed enabling technologies will in turn boost the growth of the emerging sector, thus forming a virtuous circle. As the personal computing market and the mobile computing market have matured, we examine the relation between the ecosystem market size and the semiconductor market size, for both personal computing and mobile computing. Note that for the mobile computing ecosystem, we mainly account for the market size of mobile applications; and for the personal computing ecosystem, we mainly account for the market size of software that runs on personal computers.

To project the development of AMC, we can draw parallels to the growth of the mobile computing industry: Back in the early 2000s, mobile phones had already become commodities; they were called feature phones back then. Although feature phones were almost ubiquitous, there were limited functions or applications on feature phones, and 90% of the compute power was dedicated to basic communication functions such as encoding and decoding. Since there was only less than 10% of the available compute power dedicated to applications, there were not many interesting applications on these feature phones. The whole market size of the mobile computing ecosystem was around $10 billion. Then, smartphones came to market and generated an ever-increasing demand for compute power in order to sustain an ever-increasing number of interesting mobile applications. As demand increased, a single embedded chip evolved into complex mobile SoC that consist of multicore CPUs, mobile GPUs, mobile DSPs, along with extremely good power management systems. Today, the mobile computing ecosystem enjoys a market size of $800 billion.

Currently, based on our profiling, AMC compute systems spend 95% of compute power for basic operations, such as perception, localization, and planning, leaving merely 5% for applications. This is very similar to the situation in feature phones, or the early stage of the mobile computing era. With this distribution of compute power, it is impossible to have the autonomous machines perform intelligent tasks, and thus, currently the AMC ecosystem is virtually nonexistent.

To realize the full potential of AMC, we target spending 20% of compute power on basic autonomous machine operations and leaving 80% of compute power for intelligent applications. However, to achieve this, innovations in semiconductor technologies are imperative to enable more efficient sensing, computing, and communication of AMC applications.

To fuel the rapid development of AMC, several emerging semiconductor technologies demonstrate the promise of enhancing autonomous machines' sensing, computing, and communication capabilities. Sensing is essential to any autonomous machine and the industry demands better-sensing technologies to allow autonomous machines to see further and clearer. Single-photon avalanche diode (SPAD) has the potential to capture real-time 3-D scene information while mitigating many of the downsides of existing sensors. Computing dictates the speed and duration of autonomous machines' operation. The nanosheet architecture has been proven to deliver higher compute power at the same energy consumption, or much lower energy consumption at the same compute power compared to existing FinFET technologies. Communication is fundamental for autonomous machines to cooperate. For instance, an autonomous vehicle can generate 2 GB of data per second, which even today's 5G network is not able to handle. InGaAs/InP is an upcoming technology for 6G communication, which has the potential to facilitate high-bandwidth autonomous machine communication.

10.2. Computing Solutions for Autonomous Driving

The landscape of computing solutions for autonomous driving is highly dynamic, marked by rapid advancements in semiconductor technology to meet the specific needs of intelligent vehicles. Key players in the industry like NVIDIA, MobilEye, Huawei, Horizon Robotics, and TI each contribute uniquely to this field.

NVIDIA is a major player with its Drive Thor autonomous driving system, which integrates automated driving, infotainment, and parking management. This system has been adopted by several Chinese EV manufacturers, demonstrating NVIDIA's strong presence in the market. Drive Thor is recognized for its high-performance capabilities in processing large volumes of data necessary for autonomous driving.

Mobileye, an Intel company, is a leader in autonomous driving solutions, recognized for its ADASs. Mobileye's technology is centered around its EyeQ® chip series, which powers the processing of sensory data necessary for vehicle navigation and safety features. These chips use sophisticated algorithms based on deep learning, computer vision, and sensor fusion technologies to enable capabilities such as pedestrian detection, lane recognition, and traffic sign recognition. A standout feature of Mobileye's approach is its use of a single camera for data acquisition, distinguishing it from competitors who often use a combination of cameras, radars, and LiDAR systems. This method has proven efficient in reducing costs while maintaining high levels of accuracy and reliability in real-time vehicle navigation and decision-making.

Huawei has made significant strides in the autonomous driving sector with its advanced computing solutions designed for smart vehicles. The company has developed the Huawei Mobile Data Center (MDC) platform, specifically the MDC 610, which serves as the core component in their autonomous driving systems. This platform is engineered to deliver high computing power, capable of supporting up to Level 4 autonomous driving, where vehicles can operate independently under various conditions without human intervention. A key feature of Huawei's autonomous driving solutions is their integration of both hardware and software capabilities, providing a comprehensive suite of tools for developers and manufacturers. This includes powerful AI processing capabilities, robust data security features, and efficient power management systems, making it suitable for high-performance applications in smart vehicles.

Horizon Robotics, a notable competitor from China, has developed the Journey series of processors, which are specifically tailored for autonomous driving. The Journey 5 chip, for example, is designed to handle L4 autonomous driving, indicating a high level of automation where vehicles can operate without human intervention in certain conditions. This chip is capable of performing up to 128 tera operations per second, showcasing its robust computational power. Horizon Robotics is rapidly expanding its footprint in the Chinese market, with several local car manufacturers showing keen interest in integrating these chips into their upcoming models.

TI, another significant contributor, focuses on automotive processors that drive innovation in ADASs and autonomous vehicle technologies. TI's products are designed to support a wide range of automation levels, demonstrating versatility and scalability in their application across different vehicle platforms.

As of mid-2023, NVIDIA held approximately 53% of the navigate-on-autopilot (NOA) market, underlining its strong position, particularly in high-performance sectors of the market. Mobileye continues to be a leader in ADAS technology, leveraging its expertise in camera-based systems to support readiness for Level 4 autonomy. Mobileye's systems are widely acclaimed for their precision and reliability in real-time navigation and decision-making tasks. Despite facing significant challenges due to US sanctions, Huawei's market presence is growing, particularly in China, where it leverages local partnerships and its strong telecommunications infrastructure to enhance its offerings in smart vehicle technologies. Horizon Robotics has quickly emerged as a strong competitor, particularly in the Chinese market, where it held a 31% market share in the first half of 2023. TI, although holding a smaller segment of the market at 9%, remains a crucial player due to its range of automotive chips that cater to various aspects of vehicle automation.

10.3. Computing Solutions for Cockpit Intelligence

The IEV cockpit intelligence market is driven by several key vendors, each contributing unique innovations and holding specific market shares. Vendors like NVIDIA, Qualcomm, TI, Horizon Robotics, and Mobileye are leading the way in developing integrated platforms that enhance the interface and user experience for vehicle occupants.

NVIDIA's offerings in the cockpit intelligence domain, particularly with their DRIVE IX platform, leverage AI to enable natural interaction between the vehicle and its occupants. This system integrates voice recognition, gesture recognition, and eye tracking to create a more responsive and intuitive user environment. NVIDIA's solutions are widely adopted due to their robust processing capabilities and deep learning technologies, which make them suitable for complex tasks required in intelligent cockpits.

Qualcomm is another significant player with its Snapdragon Automotive Cockpit Platforms, designed to transform in-car experiences by providing high-performance computing for infotainment systems, personalized comfort features, and onboard navigation, which are becoming standard in modern vehicles. Qualcomm's platforms support multiple high-resolution displays and provide advanced connectivity options, crucial for the next generation of smart vehicles.

TI focuses on providing a range of automotive chips that support various aspects of the intelligent cockpit, including digital instrument clusters and infotainment systems. Their products are known for reliable performance and support for advanced graphics and multimedia features, necessary for delivering rich, immersive user experiences.

Horizon Robotics, particularly strong in the Chinese market, offers the Journey series processors. These chips are tailored for automotive applications, including ADASs and intelligent cockpits, providing powerful AI computing capabilities to handle multiple complex tasks simultaneously. Mobileye contributes with their EyeQ chips, which are primarily known for ADAS functionalities but are increasingly being integrated into cockpit applications to enhance safety features through better monitoring and responsiveness to driver and environmental conditions.

Huawei has also actively expanded its role in the IEV sector, particularly focusing on the development of smart cockpit solutions. At the forefront of Huawei's offerings is the HarmonyOS-based smart cockpit, powered by Huawei's Kirin SoC solutions. The HarmonyOS cockpit is designed to enhance the in-car user experience by offering multiscreen, multiuser interaction capabilities, alongside advanced connectivity features that integrate seamlessly with Huawei's ecosystem. This solution aims to create a more interconnected and responsive environment within the vehicle, leveraging Huawei's extensive experience in telecommunications and software.

Market share dynamics reveal NVIDIA and Qualcomm as leading players due to their extensive technological advancements and broad adoption across global automotive manufacturers. Horizon Robotics is rapidly capturing market share in China, leveraging local manufacturing and integration capabilities. TI maintains a strong presence due to its reliability and comprehensive product offerings that cater to a wide range of automotive applications. Mobileye, although primarily focused on ADAS, is also making inroads into the intelligent cockpit market by integrating its technologies to improve safety and driver assistance features within the cockpit environment.

10.4. Other Semiconductor Solutions for IEVs

Besides computing solutions for autonomous driving and cockpit intelligence, other semiconductor solutions for IEVs include the following:

- **Power Management and Battery Systems**: TI and Infineon are leading providers in the power management sector, offering a range of ICs and metal-oxide-semiconductor field-effect transistor (MOSFETs) designed specifically for automotive applications. These companies produce components that enhance the efficiency and safety of battery management systems, ensuring optimal performance and extending the lifespan of the vehicle's battery.
- **Drivetrain Components:** ON Semiconductor and Wolfspeed are at the forefront of providing SiC and GaN semiconductors. These materials are known for their high efficiency and capability to withstand extreme voltages and temperatures, making them ideal for use in EV inverters. Their products allow for more compact, efficient designs that reduce energy loss and improve the range and performance of IEVs.
- **Sensors and Actuators**: STMicroelectronics and Bosch are prominent in the sensors and actuators market for IEVs. STMicroelectronics offers a wide range of sensors, including those for position, pressure, and motion detection, essential for the vehicle's interactive systems and performance monitoring. Bosch, known for its high-quality automotive components, provides various sensors that play critical roles in safety and functionality, from engine management to environmental sensing.
- **Connectivity and Safety Systems**: Qualcomm and NXP Semiconductors are leaders in providing semiconductor solutions that support connectivity and safety in IEVs. Qualcomm's technologies facilitate vehicle-to-everything (V2X) communications, essential for the integration of IEVs into smarter, safer transportation networks. NXP focuses on solutions that enhance vehicle safety, such as processors and controllers for stability control systems and ADASs.
- **Thermal Management**: Analog Devices is a key player in thermal management for IEVs. They supply a variety of temperature sensors and cooling system controllers that help manage the heat generated by batteries and electronic systems, ensuring they operate within safe, efficient parameters.

10.5. Latest Research Results

Building Computing Systems for Autonomous Driving: A few research works focus on building autonomous driving computing systems [3-7], and these works provide a comprehensive analysis of the constraints and design considerations specific to autonomous vehicles. These works introduce general models to evaluate design decisions across the vehicle's system rather than in isolation, showcasing a detailed approach to latency, energy, and cost constraints in vehicle computing systems. Key insights from these research works include the identification of performance requirements, specifically the breakdown of latency into component parts such as computation time, data transmission, mechanical response, and actual stopping time. This analysis aids in understanding how each component contributes to the vehicle's overall reaction speed and safety measures. Innovations discussed involve the integration of hardware and software in vehicle design. These platforms facilitate more efficient processing of sensory data and quicker response times, significant for safety-critical functions in autonomous vehicles. These works also explore how different computing models affect the performance and cost of the vehicle's operations, suggesting a move toward a more integrated and less compartmentalized approach in vehicle architecture.

Using FPGAs for IEV Computing: FPGAs are semiconductor devices that can be configured by a customer or a designer after manufacturing—hence "field-programmable." FPGAs can be reconfigured to perform specific tasks by loading different configuration data, which allows for adapting the hardware to new standards or to evolving computing requirements without needing to redesign or replace the hardware. This ability to adapt is crucial in the automotive sector, where evolving algorithms for tasks like image processing, data analysis, and real-time decision-making continue to advance rapidly.

FPGAs are highly efficient for parallel processing applications, which makes them ideal for the sensor fusion tasks required in autonomous driving systems. They can handle multiple data streams simultaneously from various sensors—such as LiDAR, radar, cameras, and ultrasonic sensors—processing these data in parallel to perform tasks like object detection, classification, and collision avoidance in real time.

FPGAs allow for quicker turnaround times because they can be configured to establish direct pathways for data. Unlike processors that might experience delays due to sequential processing or pipeline conflicts, FPGAs can be designed to handle specific workflows, thereby minimizing latency. This feature is critical for safety in IEVs, where response times need to be immediate to ensure effective operation and safety.

FPGAs can be more energy-efficient compared to traditional CPUs or GPUs for certain applications. Because FPGAs allow for specific customization, they can be optimized to perform the required tasks with minimal power consumption. This is especially important in battery-operated vehicles like IEVs, where power efficiency directly correlates with the vehicle's range and performance.

FPGAs are known for their robustness and suitability in harsh environments, which is a common requirement in automotive applications. Their reliability in handling vibrations, temperature variations, and other physical stressors makes them an excellent choice for deployment in vehicles.

Although the initial cost of an FPGA can be higher than a standard CPU or GPU, the long-term benefits of reusability and adaptability can make FPGAs more cost-effective over the lifecycle of a vehicle. Additionally, the ability to update the FPGA's configuration for new functionalities or improvements without hardware changes reduces the need for costly recalls or hardware upgrades.

Recent research works, including [8-15], explore the utilization of FPGAs in different IEV computing tasks, such as deep learning workload acceleration, autonomous driving perception acceleration, autonomous driving localization acceleration, and autonomous driving planning and control acceleration, demonstrating FPGAs' computing efficiency and energy efficiency superiority in many different IEV tasks.

Hardware Acceleration of Autonomous Driving Localization Tasks: Autonomous driving localization tasks are critical for ensuring that a vehicle can accurately determine its position within its environment. This process involves complex computations and real-time data processing, which are essential for safe and effective navigation. Due to the intense computational demands and the need for real-time responsiveness, these tasks often require hardware acceleration. Specifically, localization tasks in autonomous driving involve processing vast amounts of data from various sensors, including GPS, LiDAR, radar, cameras, and IMUs. Each of these sensors generates large streams of data that need to be processed simultaneously to determine the vehicle's exact position and orientation in relation to its surroundings. The complexity and volume of these data are beyond the capabilities of traditional CPU-centric architectures to process efficiently within the required time frames. Also, delays in processing localization tasks can directly impact vehicle safety and operational efficacy. Localization must occur in real time to ensure that the vehicle responds appropriately to dynamic road conditions, obstacles, and changes in the driving environment.

Recent research works, including [16-21], explore localization hardware acceleration frameworks to reduce latency and energy consumption effectively. Implementations of FPGAs demonstrate substantial improvements in the handling of autonomous tasks through hardware acceleration. These research works introduce a sophisticated localization algorithm that adapts dynamically to different operating environments, ensuring flexibility and robustness in autonomous navigation. Key results from these studies include a notable reduction in latency across various operational modes, such as registration, visual inertial odometry (VIO), and simultaneous localization and mapping (SLAM). Similarly, these research works demonstrate significant improvements in energy efficiency, with reductions up to 74% in specific scenarios, underscoring the potential of hardware acceleration in enhancing the performance and sustainability of autonomous vehicles.

Factor Graph Architecture: A recent innovative research direction for IEV computing is utilizing factor graphs for various workloads such as localization, perception, and planning. Factor graphs are a powerful tool in autonomous driving workloads due to their ability to model complex probabilistic systems, which are common in the sensory and perception tasks required for autonomous navigation. Autonomous vehicles must process and make sense of data from a variety of sensors—LiDAR, radar, cameras, and GPS, among others.

Factor graphs are capable of integrating these diverse data types into a unified probabilistic model. This integration helps in accurately estimating the state of the vehicle and its surroundings, crucial for tasks like localization and mapping. Decision-making in autonomous vehicles needs to be based on the most accurate and current information available. Factor graphs support this by allowing for efficient inference and updates as new sensor data arrive.

Driving environments are inherently uncertain, with many unpredictable elements. Factor graphs help manage this uncertainty by modeling the probability distributions of various possible states and observations. This probabilistic approach provides a more robust framework for decision-making under uncertainty compared to deterministic methods.

Recent research works have developed a factor graph hardware acceleration framework designed to generate hardware accelerators specifically tailored for optimization-based robotic applications [22-24]. These frameworks use factor graphs as a common abstraction, enabling the efficient transformation of complex robotics algorithms into a format suitable for hardware acceleration. These frameworks comprise a factor graph library, an optimizing compiler, and a hardware generation backend, ensuring that even nonexperts can easily convert high-level robotics algorithms into executable hardware instructions. Overall, these factor graph acceleration frameworks address the significant nonrecurring engineering (NRE) costs and inefficiencies found in existing accelerator designs by providing a scalable, low-cost solution that does not compromise on performance or efficiency.

Deep Learning Compression: Studies [25-28] focus on enabling computationally demanding deep learning workloads on edge devices through compression. Compression techniques such as pruning, quantization, and knowledge distillation are utilized to reduce the model size and computational demands without significantly compromising accuracy. Pruning involves removing redundant or less important weights from a model, effectively reducing the number of parameters and operations required during inference. Quantization reduces the precision of the numerical values in the model, allowing operations to be executed more rapidly, often with specialized hardware support for low-precision arithmetic. Knowledge distillation involves training a smaller, more compact model (the student) to mimic the behavior of a larger, pretrained model (the teacher), enabling the smaller model to achieve performance levels close to those of the larger model but with fewer computational resources.

Middleware Innovations: Besides developing innovative semiconductor solutions for IEVs, many research works focus on hardware–software co-optimizations. Studies [29-33] focus on middleware innovations in IEV applications. Middleware in the context of autonomous driving serves as the crucial software layer that connects the complex hardware and software components of autonomous vehicle systems. Its primary function is to manage communication and data flow between different parts of the system, ensuring that higher-level applications can operate effectively without needing to manage low-level processes directly.

Middleware abstracts the complexities of the underlying hardware from the application layers. This allows developers to focus on improving autonomous driving algorithms without worrying about the specifics of hardware interfaces. By providing a common programming interface, middleware makes software development more efficient and portable across different hardware platforms.

Middleware handles communication both internally (between components of the car) and externally (with cloud services or other vehicles in V2X scenarios). This includes managing data transmission over various networks and protocols and ensuring timely and reliable data exchange, which is crucial for real-time decision-making processes in autonomous driving.

Middleware efficiently manages the data flow, prioritizing and routing sensor data from devices like LiDAR, cameras, and radars to processing units and storage systems. It ensures that data integrity is maintained and that the data lifecycle is managed efficiently, from creation to deletion.

Middleware facilitates the integration of various software and hardware components, ensuring they work together seamlessly. It supports interoperability, which is essential in systems where components come from different vendors, each with its own standards and specifications.

Middleware optimizes resource allocation, ensuring that computational and memory resources are utilized efficiently to handle the high workloads of autonomous driving tasks. Middleware can monitor system performance and dynamically allocate resources to different tasks based on priority and current system load, enhancing overall system efficiency.

10.6. Summary

A central theme of this chapter is the concept of a "virtuous cycle" of technological and economic growth driven by semiconductor innovations. This cycle illustrates how advances in semiconductor technology not only support but amplify the market size of their ecosystems, substantially outpacing the growth of the semiconductor industry itself. Historical data comparing the market sizes of personal and mobile computing ecosystems to their semiconductor markets highlight a potential for the semiconductor industry to catalyze a 15- to 25-fold increase in the market size of the AMC sector. Under this central theme, we present the existing semiconductor solutions for autonomous driving, cockpit intelligence, and other IEV usage scenarios. We have presented a review of the latest research results on semiconductor innovations for IEVs.

References

1. Liu, S., "Shaping the Outlook for the Autonomy Economy," *Communications of the ACM* 67, no. 6 (2024): 10-12.
2. Liu, S. and Gaudiot, J.L., "IEEE International Roadmap for Devices and Systems (IRDS) Autonomous Machine Computing," White Paper, 2023.
3. Liu, S., Tang, J., Zhang, Z., and Gaudiot, J.L., "Computer Architectures for Autonomous Driving," *Computer* 50, no. 8 (2017): 18-25.
4. Liu, S., Liu, L., Tang, J., Yu, B. et al., "Edge Computing for Autonomous Driving: Opportunities and Challenges," *Proceedings of the IEEE* 107, no. 8 (2019): 1697-1716.
5. Yu, B., Hu, W., Xu, L., Tang, J. et al., "Building the Computing System for Autonomous Micromobility Vehicles: Design Constraints and Architectural Optimizations," in *2020 53rd Annual IEEE/ACM International Symposium on Microarchitecture (MICRO)*, Athens, Greece, 2020, 1067-1081, IEEE.

6. Yu, B., Tang, J., and Liu, S., "On Designing Computing Systems for Autonomous Vehicles: A PerceptIn Case Study," in *Proceedings of the 26th Asia and South Pacific Design Automation Conference*, Tokyo, Japan, 2021, 742-747.

7. Fang, W., Zhang, Y., Yu, B., and Liu, S., "DragonFly+: FPGA-Based Quad-Camera Visual SLAM System for Autonomous Vehicles," in *Proceedings IEEE HotChips*, Cupertino, CA, 2018, 1.

8. Liu, S., Wan, Z., Yu, B., and Wang, Y., *Robotic Computing on FPGAs* (San Francisco: Morgan & Claypool Publishers, 2021).

9. Wan, Z., Yu, B., Li, T.Y., Tang, J. et al., "A Survey of FPGA-Based Robotic Computing," *IEEE Circuits and Systems Magazine* 21, no. 2 (2021): 48-74.

10. Liu, Q., Hao, Y., Liu, W., Yu, B. et al., "An Energy Efficient and Runtime Reconfigurable Accelerator for Robotic Localization," *IEEE Transactions on Computers* 72, no. 7 (2022): 1943-1957.

11. Liu, Q., Wan, Z., Yu, B., Liu, W. et al., "An Energy-Efficient and Runtime-Reconfigurable FPGA-Based Accelerator for Robotic Localization Systems," in *2022 IEEE Custom Integrated Circuits Conference (CICC)*, Newport Beach, 2022, 1-2, IEEE.

12. Wan, Z., Lele, A., Yu, B., Liu, S. et al., "Robotic Computing on FPGAs: Current Progress, Research Challenges, and Opportunities," in *2022 IEEE 4th International Conference on Artificial Intelligence Circuits and Systems (AICAS)*, Incheon, Republic of Korea, 2022, 291-295, IEEE.

13. Gao, T., Wan, Z., Zhang, Y., Yu, B. et al., "IELAS: An ELAS-Based Energy-Efficient Accelerator for Real-Time Stereo Matching on FPGA Platform," in *2021 IEEE 3rd International Conference on Artificial Intelligence Circuits and Systems (AICAS)*, Washington, DC, 2021, 1-4, IEEE.

14. Wan, Z., Zhang, Y., Raychowdhury, A., Yu, B. et al., "An Energy-Efficient Quad-Camera Visual System for Autonomous Machines on FPGA Platform," in *2021 IEEE 3rd International Conference on Artificial Intelligence Circuits and Systems (AICAS)*, Washington, DC, 2021, 1-4, IEEE.

15. Fang, W., Zhang, Y., Yu, B., and Liu, S., "FPGA-Based ORB Feature Extraction for Real-Time Visual SLAM," in *2017 International Conference on Field Programmable Technology (ICFPT)*, Melbourne, VIC, Australia, 2017, 275-278, IEEE.

16. Liu, W., Yu, B., Gan, Y., Liu, Q. et al., "Archytas: A Framework for Synthesizing and Dynamically Optimizing Accelerators for Robotic Localization," in *MICRO-54: 54th Annual IEEE/ACM International Symposium on Microarchitecture*, Athens, Greece, 2021, 479-493.

17. Gan, Y., Bo, Y., Tian, B., Xu, L. et al., "Eudoxus: Characterizing and Accelerating Localization in Autonomous Machines Industry Track Paper," in *2021 IEEE International Symposium on High-Performance Computer Architecture (HPCA)*, Seoul, Korea, 2021, 827-840, IEEE.

18. Liu, Q., Qin, S., Yu, B., Tang, J. et al., "π-BA: Bundle Adjustment Hardware Accelerator Based on Distribution of 3D-Point Observations," *IEEE Transactions on Computers* 69, no. 7 (2020): 1083-1095.

19. Qin, S., Liu, Q., Yu, B., and Liu, S., "π-BA: Bundle Adjustment Acceleration on Embedded FPGAs with Co-Observation Optimization," in *2019 IEEE 27th Annual International Symposium on Field-Programmable Custom Computing Machines (FCCM)*, San Diego, 2019, 100-108, IEEE.

20. Tang, J., Yu, B., Liu, S., Zhang, Z. et al., "π-SOC: Heterogeneous SOC Architecture for Visual Inertial Slam Applications," in *2018 IEEE/RSJ International Conference on Intelligent Robots and Systems (IROS)*, Madrid, Spain, 2018, 8302-8307, IEEE.

21. Zhang, Z., Liu, S., Tsai, G., Hu, H. et al., "PIRVS: An Advanced Visual-Inertial Slam System with Flexible Sensor Fusion and Hardware Co-Design," in *2018 IEEE International Conference on Robotics and Automation (ICRA)*, Brisbane, QLD, Australia, 2018, 3826-3832, IEEE.

22. Hao, Y., Gan, Y., Yu, B., Liu, Q. et al., "ORIANNA: An Accelerator Generation Framework for Optimization-Based Robotic Applications," in *Proceedings of the 29th ACM International Conference on Architectural Support for Programming Languages and Operating Systems*, Vol. 2, San Diego, 2024, 813-829.

23. Hao, Y., Gan, Y., Yu, B., Liu, Q. et al., "BLITZCRANK: Factor Graph Accelerator for Motion Planning," in *2023 60th ACM/IEEE Design Automation Conference (DAC)*, San Francisco, 2023, 1-6, IEEE.

24. Hao, Y., Yu, B., Liu, Q., Liu, S. et al., "Factor Graph Accelerator for LiDAR-Inertial Odometry," in *Proceedings of the 41st IEEE/ACM International Conference on Computer-Aided Design*, San Diego, 2022, 1-7.

25. Guan, H., Liu, S., Ma, X., Niu, W. et al., "CoCoPIE: Enabling Real-Time AI on Off-the-Shelf Mobile Devices via Compression-Compilation Co-Design," *Communications of the ACM* 64, no. 6 (2021): 62-68.

26. Sung, H.H., Xu, Y., Guan, J., Niu, W. et al., "Brief Industry Paper: Enabling Level-4 Autonomous Driving on a Single $1k Off-the-Shelf Card," in *2022 IEEE 28th*

Real-Time and Embedded Technology and Applications Symposium (RTAS), Milano, Italy, 2022, 297-300, IEEE.

27. Zhao, P., Niu, W., Yuan, G., Cai, Y. et al., "Brief Industry Paper: Towards Real-Time 3D Object Detection for Autonomous Vehicles with Pruning Search," in *2021 IEEE 27th Real-Time and Embedded Technology and Applications Symposium (RTAS)*, Nashville, 2021, 425-428, IEEE.

28. Zhang, K., Liu, Y., Zhang, W., and Liu, S., "π-Learner: A Lifelong Roadside Learning Framework for Infrastructure Augmented Autonomous Driving," *Computer* 55, no. 6 (2022): 30-39.

29. Liu, L., Tang, J., Liu, S., Yu, B. et al., "π-rt: A Runtime Framework to Enable Energy-Efficient Real-Time Robotic Vision Applications on Heterogeneous Architectures," *Computer* 54, no. 4 (2021): 14-25.

30. Li, D., Tang, J., and Liu, S., Brief Industry Paper: An Edge-Based High-Definition Map Crowdsourcing Task Distribution Framework for Autonomous Driving," in *2021 IEEE 27th Real-Time and Embedded Technology and Applications Symposium (RTAS)*, Nashville, 2021, 453-456, IEEE.

31. Xie, J., Tang, J., Wang, Y., Zhu, Q. et al., "Brief Industry Paper: An Infrastructure-Aided High Definition Map Data Provisioning Service for Autonomous Driving," in *2021 IEEE 27th Real-Time and Embedded Technology and Applications Symposium (RTAS)*, Nashville, 2021, 421-424, IEEE.

32. Liu, S., Wang, J., Wang, Z., Yu, B. et al., "Brief Industry Paper: The Necessity of Adaptive Data Fusion in Infrastructure-Augmented Autonomous Driving System," in *2022 IEEE 28th Real-Time and Embedded Technology and Applications Symposium (RTAS)*, Milano, Italy, 2022, 293-296, IEEE.

33. Liu, S., Yu, B., Liu, Y., Zhang, K. et al., "Brief Industry Paper: The Matter of Time—A General and Efficient System for Precise Sensor Synchronization in Robotic Computing," in *2021 IEEE 27th Real-Time and Embedded Technology and Applications Symposium (RTAS)*, Nashville, 2021, 413-416, IEEE.

Chapter 11

Connectivity for Intelligent Electric Vehicles

Connected vehicles are not just automobiles with technology; they are the epitome of a new era where transportation converges with communication, redefining the very essence of mobility and connectivity.

—Anonymous

Connectivity is a cornerstone for the development and functionality of IEVs, serving as the backbone for a multitude of features that enhance safety, efficiency, and user experience. Connectivity is crucial to IEVs for the following reasons:

- **Real-time data processing and sharing:** Connectivity enables IEVs to process and share data in real time with other vehicles, infrastructure (like traffic lights and road signs), and cloud-based services. This Vehicle-to-Everything (V2X) communication is essential for optimizing traffic flow, reducing congestion, and enhancing safety by preventing accidents. Real-time data sharing allows IEVs to make informed decisions, adjust to changing road conditions, and navigate more efficiently [1, 2].
- **Autonomous driving:** For autonomous or semiautonomous vehicles, connectivity is vital for the vehicles' ability to make decisions. By accessing real-time traffic data, weather conditions, and updates from other vehicles, IEVs can adjust their driving strategy to enhance safety and efficiency [3, 4].
- **Remote monitoring and diagnostics:** Connectivity allows for the remote monitoring and diagnostics of vehicles, enabling manufacturers and service providers to access vehicle data in real time. This can lead to predictive

maintenance, where potential issues are identified and addressed before they lead to breakdowns, significantly improving reliability and reducing maintenance costs.

- **Enhanced user experience:** Connectivity offers IEV users access to a wide range of services and features, from real-time navigation and traffic updates from HD maps to on-demand entertainment and vehicle updates [5, 6]. This not only improves the driving experience but also enables the vehicle to stay up-to-date with the latest software, much like a smartphone.
- **Energy management and efficiency:** For EVs, in particular, connectivity plays a crucial role in energy management. It allows vehicles to identify the nearest charging stations, optimize charging times based on energy costs and grid demand, and integrate with smart home systems for seamless energy management.

Connectivity transcends being merely an optional enhancement for IEVs; it constitutes a fundamental pillar vital for their functionality, safety, and attractiveness to users. With the continuous evolution of technology, the significance and scope of connectivity in molding the future landscape of IEVs are poised for expansion. In this chapter, focusing on the usage scenario of IVCAD, we delve into the prevailing communication challenges faced by IEVs, exploring potential solutions and strategies to overcome these hurdles [7].

11.1. Infrastructure Vehicle Cooperative Autonomous Driving

The most popular connectivity usage scenario is IVCAD, which relies on roadside perception to significantly improve the safety of autonomous vehicles compared to the on-vehicle-only autonomous driving approach. Specifically, IVCAD extends the perception range from 70 m (with the on-vehicle perception only) to over 1000 m (by fusing perception results from multiple roadside perception units within a region; hence, the whole system knows more of the environment).

In addition, the disengagement rate measures how often a safety driver has to take over control of the autonomous vehicle in cases that the autonomous driving system cannot handle. With the deployment of IVCAD, the disengagement rate has been reduced by more than 90% in multiple busy intersections in our deployment where the on-vehicle-only perception fails to detect the blind spots. However, network communication is a major technical roadblock for reliable IVCAD deployments.

On-vehicle-only autonomous driving systems usually rely on raw sensing data from mmWave radars, LiDARs, cameras, and Global Navigation Satellite System (GNSS)/IMUs, and each sensor produces raw data at a different frequency. The processing components are invoked with different frequencies, performing computation using the latest input data and producing outputs to downstream components periodically. The cameras capture images at 30 FPS and feed the raw data to the 2-D perception module, the LiDARs capture point clouds at 10 FPS and feed the raw data to the 3-D perception module as well as the localization module, the GNSS/IMUs generate positional updates at 100 Hz and feed the raw data to the localization module, and the mmWave radars detect obstacles at 10 FPS and feed the raw data to the perception fusion module.

The results of 2-D and 3-D perception are fed into the perception fusion module to create a comprehensive perception list of all detected objects.

The perception list is then fed into the tracking module to create a tracking list of all detected objects. The tracking list then is fed into the prediction module to create a prediction list of all objects. After that, both the prediction results and the localization results are fed into the planning module to generate a navigation plan. The navigation plan then is fed into the control module to generate control commands, which are finally sent to the autonomous vehicle for execution at 100 Hz.

The on-vehicle-only autonomous driving approach, while widely experimented and adopted by many companies, faces fundamental reliability, safety, and accuracy challenges going forward. In detail, on-vehicle sensors such as the cameras and LiDARs have limited detection ranges and cannot handle occlusion problems. Worse, cameras are challenged by varying lighting conditions, and LiDARs can be confused by varying weather conditions such as fog and snowfall. Unusual, unfamiliar, and unstructured situations, such as accidents and road work, can be hard to classify. Even if on-vehicle sensors could capture perfect information, the compute capability of a single vehicle would unlikely be sufficient for real-time processing as algorithms become increasingly complicated.

The limitations of on-vehicle-only autonomous vehicles can be overcome by vehicle-to-everything (V2X) communication technology. With V2X communication, cooperative autonomous driving could be achieved by obtaining augmented sensing and maneuvering information in real time from surrounding vehicles and infrastructures, which will enhance the safety, reliability, and efficiency of autonomous driving. IVCAD combines intelligent infrastructures, dubbed SoRs and autonomous vehicles, dubbed SoVs, along with the ITCS into one integral and fully intelligent transportation system.

Specifically, the SoV consists of an Industrial PC (IPC), a series of sensors, as well as the ECUs and the CAN bus connecting all components (the ECUs and the CAN bus are standard autonomous vehicle components). The SoR consists of similar but more advanced equipment. Different from the SoV, the SoR's IPC is configured with a high-end Nvidia 1080Ti GPU along with other sensors. The interactions between the SoVs and the SoRs are highly latency-sensitive. Each SoR is equipped with a RSU, and each SoV is equipped with an OBU. The RSUs and OBUs communicate directly to facilitate the fusion of the SoR data and the SoV data on the vehicle side.

Figure 11.1 presents a detailed end-to-end illustration of our cooperative autonomous driving system. Starting from the left side, the sensing systems of the SoV and the SoR generate raw sensing data from mmWave radars, LiDARs, cameras, GNSS receivers, and IMUs, where each sensor produces raw data at its own frequency. The cameras capture images at 30 FPS, the LiDARs capture point clouds at 10 FPS, and the GNSS/IMUs generate positional updates at 100 Hz.

Next, the 2-D perception node consumes raw images for object detection and scene segmentation. The 3-D perception node consumes raw LiDAR scans for object shape and type detection. The perception fusion node consumes outputs from the 2-D perception node, the 3-D perception node, as well as the mmWave radars raw data to create a comprehensive perception list of all detected objects. The localization node is only needed on the SoVs. It consumes and fuses raw LiDAR scans as well as GNSS/IMU data for vehicle positional updates. Then, the

Figure 11.1 Infrastructure–vehicle cooperative autonomous driving (IVCAD).

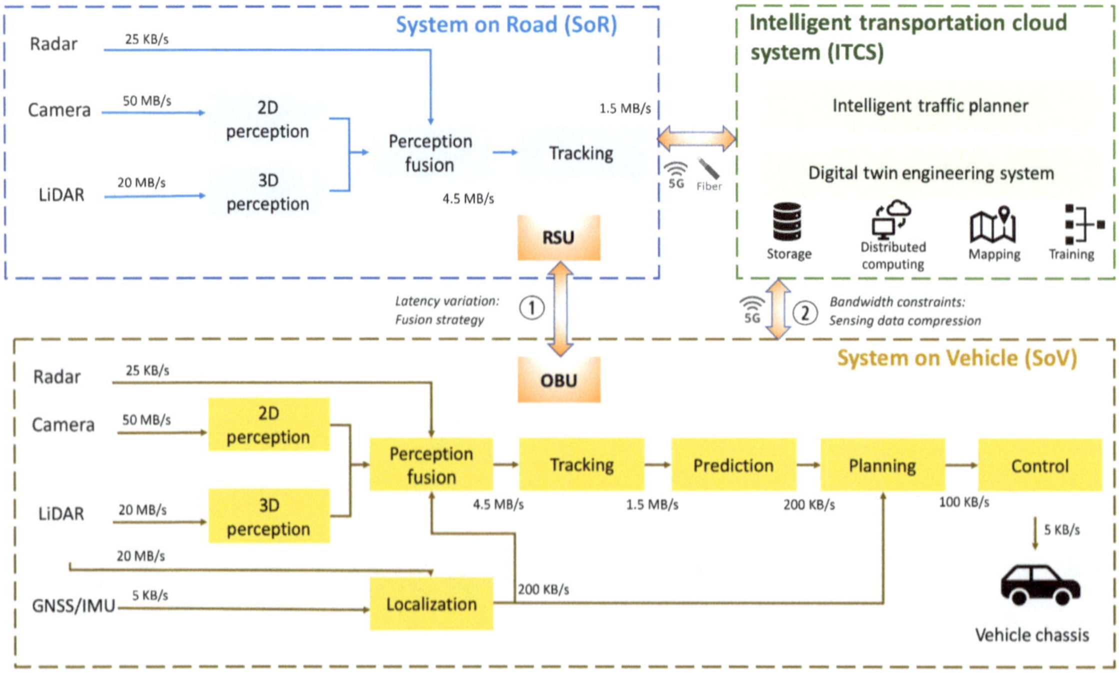

interaction between the SoRs and the SoVs starts: the output of the SoR's tracking node is fed to the SoV's tracking node, which generates a comprehensive perception list as well as the historical trajectory of each detected object. Note that the interaction between the SoRs and the SoVs is highly sensitive to latency jitters. To ensure the reliability and safety of the vehicles, the SoV is designed to tolerate up to 50 ms of latency delay from the SoR.

At last, the SoV has comprehensive tracking data, which are fed to the prediction node to generate the prediction list. Both the prediction list and the localization results are fed into the planning node at 10 Hz to generate a navigation plan. The navigation plan then is fed into the control node at 10 Hz to generate control commands, which are finally sent to the autonomous machine for execution at 100 Hz. In addition to the latency-sensitive SoV–SoR interactions, the SoRs and the SoVs also communicate directly with the ITCS. The SoRs and the SoVs will upload the raw sensing data for cloud-based applications such as simulation, deep learning model training, and HD map generation.

The uploading of sensing data can tolerate latency variation but is bandwidth-bound. It demands at least 25 MB/s uplink bandwidth for streaming essential raw data to the ITCS. Three modes of communication are supported: C-V2X, commercial 5G network, and optical fiber network. The latency-sensitive communication between the SoRs and the SoVs can be done through C-V2X and commercial 5G network; the bandwidth-constrained communication between the SoVs and the ITCS can be done through commercial 5G network; and the communication between the SoRs and the ITCS can be done through commercial 5G network or optical fiber network. We never experienced any bandwidth or latency problems with the optical fiber network; nonetheless, the C-V2X network does not provide sufficient bandwidth and the commercial 5G network contains too many latency jitters for the cooperative autonomous driving workloads.

11.2. Real-World Communication Challenges

In this section, we delve into the details of the communication challenges we have observed in our deployments: First, the current mobile network bandwidth is constraining the uploading of raw sensing data, which is crucial for cloud-based applications such as deep learning model training and HD map generation.

Second, the network latency jitters remain high for a considerable portion of a vehicle's trip, greatly impacting the reliability and safety of autonomous vehicles. Note that in our deployments, LTE V2X is used for broadcasting simple messages of vehicle states and 5G mobile network is used for transferring the large volume of perception data. While 3GPP (3rd Generation Partnership Project) has recently completed 5G-NR release 16 to enhance V2X direct communication, the state-of-the-art commodity OBU and RSU only support LTE V2X direct communication (specified by 3GPP release 14), of which communication capacity is insufficient for infrastructure-based perception.

Our field testing, in which ZTE's and Huawei's RSUs and OBUs are used to perform direct communication via the PC5 interface, verifies that although the latency of LTE direct communication (20 to 30 ms) is sufficient for fusing roadside tracking results on the vehicle side, the maximum bandwidth (200 kB/s) is far below the required data transmitting rate of the roadside infrastructure, even when only transmitting the results of the tracking module demands 1.5 MB/s. Because of the limited bandwidth, direct communication is used in our real deployment to broadcast small-sized packages for some latency-sensitive use cases, such as platooning.

In our deployment, a commercial 5G mobile network is used for providing sufficient bandwidth and transferring roadside tracking results to the vehicle in real time. In terms of bandwidth, the commercial 5G network can achieve 25 MB/s throughput for downloads, which is sufficient for fusing roadside perception results (1.5 MB/s) on the vehicle, and provides 9 MB/s throughput for uploads, which is sufficient for uploading autonomous vehicle's ego states.

To have a comprehensive understanding of the wireless communication network environment, we also tried with the commercial 4G LTE network. The test results indicate that the 4G LTE network provides 4 MB/s downlink bandwidth and 2.5 MB/s uplink bandwidth, which can satisfy the bandwidth requirements of the infrastructure-assisted fusion system, but is far from meeting the bandwidth requirement of sending sensor data.

We conducted 4G and 5G latency measurements for several rounds on our deployment site, in which data from SoR are first transferred to a local edge server through direct fiber connections, then transferred to core networking using TCP/IP protocol, and finally sent to SoV through mobile network. Derived from experimental measurements, the 5G network has an average latency of 30 ms, which satisfies our requirements. However, latency jitters can be as large as 100 ms, which leads to serious frame dropping while performing the infrastructure–vehicle fusion operations.

The 4G LTE network exhibits rather long latency and large jitters, which cannot be used for fusing roadside information timely.

In summary, disappointingly, the field experimental results demonstrate that real-world wireless networks do not simultaneously provide low latency, high bandwidth, and low jitter, which significantly hurts the timely arrivals of the infrastructure-side data and the uploading of raw sensing data. While we have little control over the commercially deployed communication networks, in the next two sections, we delve into two techniques: one to cope with the bandwidth limitation problem to facilitate continuous sensing data uploading, and the other to cope with the latency jitter problems to enable reliable real-time infrastructure–vehicle cooperation.

11.3. Overcoming Bandwidth Limitation with Data Compression

In this section, we explain how we address the bandwidth limitation challenge with the sensing data compression strategy [8]. We upload raw sensing data, including images and LiDAR-generated point clouds, which are used for HD map creation, deep learning model training, and simulation workloads. In particular, point cloud data require continuous transmission to assist HD map construction; point cloud data are large in size and thus challenge the bandwidth limit of the communication technologies.

Today's LiDARs generate hundreds of thousands of points each frame, amounting to up to 25 MB of raw data per second, which leads to about 1 TB of data assuming 10 h of operation time per vehicle. Worse, an HD map is typically updated at least weekly, thus requiring continuous, as opposed to one-off, collecting point cloud data from the vehicles. However, vehicles have only cellular accesses, part of whose bandwidth has to be used to communicate with the RSUs (and potentially other vehicles in the future). With more autonomous vehicles on the road in the same locality, point cloud data transmission will undoubtedly congest the base stations. Hence, our main challenge is to compress at least 25 MB/s of raw LiDAR data with only 9 MB/s of uplink bandwidth available.

Point cloud compression must also be done in real time, that is, match the data generation rate, in order to ease local storage pressure. The main idea of our compression system is to exploit redundancies both within a point cloud (spatial) and across point clouds (temporal). Spatially, many surfaces in the real world are planes (e.g., walls and ground); even nonplane surfaces could be approximated by a set of planes. Temporally, consecutive point clouds share a great chunk of overlapped areas of the scene; thus, the same set of planes could be used to encode points across point clouds. While intuitive, exploiting spatial and temporal redundancies in real time is challenging due to the irregular/unstructured point cloud and the compute-intensive plane fitting process.

Our system consists of three main blocks: range image conversion, spatial encoding, and temporal encoding. Given a sequence of consecutive point clouds, we differentiate between two point cloud types: key point cloud (K-frame) and predicted point cloud (P-frame). A sequence has only one K-frame and the rest is P-frames. P-frames are transformed (both translation and rotation) to K-frame's coordinate system using the IMU measurements.

After transformation, each point cloud is converted to a range image for subsequent computations. The range image not only provides an initial compression to the original point cloud, but also provides a structured representation of the (unstructured) point cloud that is hardware-friendly. We then spatially encode K-frame by fitting planes; the fitted planes in the K-frame are then (re-)used to temporally encode P-frames, greatly improving the overall compression rate and speed.

To be robust against transformation errors, which might be introduced due to noisy IMU observations, we propose a set of techniques that compensate for the sensor noise and preserve the encoding quality. In the end, after spatial and temporal encoding, most of the tiles in the range images are plane-encoded; the unfit tiles are left in what we call residual maps. The planes and the residual maps are then further compressed by a lossless compression scheme (e.g., Huffman encoding) to generate the final encoded data.

Overall, our compression method achieves up to 90 times the compression rate, significantly outperforming MPEG's LiDAR point cloud compression standard. The compressed point clouds also provide a new avenue for video compression through signal reconstruction and enhancement. In particular, sparse point clouds can be seen as possessing accurate depth information about key points in an image. In this way, the vehicles can afford to capture videos in a much lower resolution and at a much lower rate, relying on the point cloud-assisted enhancement pass to reconstruct the original video signals.

11.4. Overcoming Latency Variation with Adaptive Fusion

In this section, we explain how we address the latency variation challenge with the adaptive fusion strategy. After receiving the SoR's tracking data, the SoV's OBU will fuse the data with its own tracking module's output. We call this process tracking fusion [3]. Because of the variation in network latency, SoR's data often arrive nondeterministically, deviating from the expected arrival time by as much as 100 ms. However, there is only a very small time window (in our case 50 ms) that the SoV could wait for the SoR data arrival, since the entire system must operate in real time to ensure safety. Thus, if the SoV cannot receive the data from the SoR on time, the SoV has no choice but to skip the tracking fusion process; instead, the vehicle would have to track other vehicles without the global information from the SoR, potentially degrading the task accuracy and, by extension, safety.

In our field tests, we find that, in the worst case, consecutive frames from the SoR are discarded by the SoV. As a result, we observed a deadline miss ratio as high as 8% of the tracking fusion task, greatly impacting the reliability and safety of autonomous vehicles. One way to address the latency jitter issue is if the semantics of the communicated data is inherently tolerant to delay. In particular, we could offload the tracking task to the SoR (which does not add much load to the SoR given that the SoR is equipped with a more powerful computing unit compared to the SoV). An implication of offloading tracking to SoR is that trajectory prediction will have to be offloaded to the SoR as well, since prediction depends on the tracking data.

We could then communicate the prediction results from the SoR to the SoV. We call this process prediction fusion, as opposed to tracking fusion, since it is the prediction results that get fused. Prediction fusion is less sensitive to network latency jitters and more tolerant to network congestion because of two key reasons. First, the amount of data required for prediction fusion (200 kB/s) is less than that of tracking fusion (1.5 MB/s). Second, the semantics of prediction results dictates that they are more tolerant to network delay, because each prediction result contains the predicted trajectory of a few seconds into the future (5 sec in our case). Even if one prediction packet is dropped or arrives late, the previous prediction packet still has sufficient information in the predicted trajectory to cover a 5-sec temporal span.

However, fusing prediction results introduces more noise for the planning module compared to fusing the tracking results. Assuming the current time is T, compared to the actual trajectory at T+1 that should have arrived but did not, we now resort to the predicted trajectory at T+1 (from the predicted packet), which is less accurate. The more prediction packet drops, the less accurate our downstream algorithm will be. This trade-off between response time and accuracy motivated us to develop an adaptive fusion engine to dynamically choose a fusion method to achieve the best SoV performance under different situations.

The adaptive fusion engine monitors the network condition in real time, measured by the number of dropped SoR fusion frames within a sliding window. If the number of dropped frames crosses a threshold, the adaptive fusion engine selects the planning fusion method; otherwise, it stays with the tracking fusion method. To implement adaptive fusion, we developed a host component on the SoV side (OBU) and an agent component on the SoR side (RSU). The RSU and the OBU use almost the same software stack, including the PHY layer, MAC layer, IP layer, and control layer. We added an adaptive fusion layer above the control layer; the OBU monitors the network conditions on the SoV side and makes adaptive fusion decisions accordingly. The OBU notifies the RSU as to whether to switch to a different fusion approach to improve the fusion efficiency. The field test results verify that the adaptive fusion method has significantly improved autonomous driving's safety and reliability by decreasing the fusion miss ratio from 8% to less than 1%.

11.5. Summary

Connectivity stands as a crucial element for IEVs, yet the reality of current deployments presents significant challenges. The limited bandwidth of networks restricts the upload of raw sensor data, while the variability in real-world network latency poses serious concerns for the reliability and safety of autonomous vehicles.

Despite these challenges and our constrained influence over commercial wireless communication networks, we have innovated and validated two practical solutions. The first is a sensing compression strategy designed to alleviate the bandwidth constraints, enabling more efficient data transmission. The second solution is an adaptive fusion engine tailored to manage the variability in network latency, ensuring consistent and reliable data integration for autonomous vehicle operation. It is important to highlight that these solutions, while implemented in our systems, possess a universal applicability across any IVCAD system.

The overarching aim of this chapter is to catalyze the advancement of communication networks that guarantee ample bandwidth and minimal latency variations, which are essential for the seamless operation of autonomous vehicles. To this end, we outline potential research directions. One approach involves negotiating with commercial wireless networks to prioritize connections for autonomous machines, given their critical role in safety. Additionally, the concept of a virtual communication backbone emerges as a promising solution. By leveraging existing LTE or 5G networks in tandem with direct C-V2X communication, this backbone could significantly enhance the reliability of communications for autonomous machines through strategic scheduling.

This discourse not only sheds light on the current connectivity hurdles faced by IEVs but also proposes forward-thinking solutions and research avenues to overcome these obstacles, paving the way for a future where autonomous vehicles can operate with unparalleled reliability and efficiency.

References

1. Liu, S., Yu, B., Tang, J., and Zhu, Q., "Towards Fully Intelligent Transportation through Infrastructure-Vehicle Cooperative Autonomous Driving: Challenges and Opportunities," in *2021 58th ACM/IEEE Design Automation Conference (DAC)*, San Francisco, 2021, 1323-1326, IEEE.
2. Zhang, K., Liu, Y., Zhang, W., and Liu, S., "π-Learner: A Lifelong Roadside Learning Framework for Infrastructure Augmented Autonomous Driving," *Computer* 55, no. 6 (2022): 30-39.
3. Liu, S., Wang, J., Wang, Z., Yu, B. et al., "Brief Industry Paper: The Necessity of Adaptive Data Fusion in Infrastructure-Augmented Autonomous Driving System," in *2022 IEEE 28th Real-Time and Embedded Technology and Applications Symposium (RTAS)*, Milano, Italy, 2022, 293-296, IEEE.
4. Yao, L., Zhao, H., Tang, J., Liu, S. et al., "Streaming Data Priority Scheduling Framework for Autonomous Driving by Edge," in *2021 IEEE 45th Annual Computers, Software, and Applications Conference (COMPSAC)*, Madrid, Spain, 2021, 37-42, IEEE.
5. Li, D., Tang, J., and Liu, S., "Brief Industry Paper: An Edge-Based High-Definition Map Crowdsourcing Task Distribution Framework for Autonomous Driving," in *2021 IEEE 27th Real-Time and Embedded Technology and Applications Symposium (RTAS)*, Nashville, 2021, 453-456, IEEE.
6. Xie, J., Tang, J., Wang, Y., Zhu, Q. et al. "Brief Industry Paper: An Infrastructure-Aided High Definition Map Data Provisioning Service for Autonomous Driving," in *2021 IEEE 27th Real-Time and Embedded Technology and Applications Symposium (RTAS)*, Nashville, 2021, 421-424, IEEE.
7. Liu, S., Yu, B., Tang, J., Zhu, Y. et al., "Communication Challenges in Infrastructure-Vehicle Cooperative Autonomous Driving: A Field Deployment Perspective," *IEEE Wireless Communications* 29, no. 4 (2022): 126-131.
8. Feng, Y., Liu, S., and Zhu, Y., "Real-Time Spatio-Temporal Lidar Point Cloud Compression," in *2020 IEEE/RSJ International Conference on Intelligent Robots and Systems (IROS)*, Las Vegas, 2020, 10766-10773, IEEE.

Chapter 12

Safety Standards for Intelligent Electric Vehicles

Quality means doing it right when no one is looking.

—Henry Ford

Automotive safety standards are regulations and guidelines designed to ensure the safety and reliability of motor vehicles. They cover a broad range of components and systems within a vehicle, including structural integrity, occupant protection, and road handling. These standards are set by government agencies, international bodies, and industry groups to mitigate risks associated with vehicle operation, reduce the likelihood of accidents, and enhance the overall safety of passengers, pedestrians, and other road users. In this section, we examine the safety standards for IEVs.

12.1. Overview of Safety Standards for IEVs

Safety standards encompass various aspects of vehicle design and performance, such as crash-worthiness, which involves a vehicle's ability to absorb impact and protect occupants during a collision. They also cover the performance of safety features like airbags, seat belts, and braking systems. Additionally, as technology evolves, these standards increasingly address modern features like electronic stability control, autonomous driving aids, and cybersecurity measures.

By setting minimum safety criteria that vehicles must meet before they can be sold or operated on public roads, automotive safety standards play a crucial role in reducing fatalities and injuries resulting from road accidents. Compliance with these standards is mandatory for manufacturers, and adherence is monitored through testing and certification processes.

Safety standards are essential in IEVs for several reasons, fundamentally to ensure the safety and reliability of these technologically advanced vehicles. As IEVs incorporate complex electrical systems, sophisticated software, and advanced connectivity features, they introduce new kinds of risks that traditional vehicles do not face. Safety standards help mitigate these risks by establishing guidelines that manufacturers must follow to protect users and the public.

First, safety standards address the specific challenges posed by the high-voltage batteries used in IEVs. These batteries, typically lithium-ion, are prone to thermal runaway risks if damaged or improperly managed, potentially leading to fires or explosions. Standards ensure that robust battery management systems are in place to monitor battery health, control charging and discharging rates, and manage thermal conditions to prevent such incidents. They also set out testing protocols for crash resilience, ensuring that the battery's integrity is maintained even in severe collisions.

Moreover, the autonomy aspect of IEVs, which may range from partial to full driving automation, requires rigorous safety protocols to ensure that autonomous systems can safely handle driving tasks. This involves not just the physical safety of passengers and pedestrians but also the functional safety of vehicular software and hardware. Standards like ISO 26262 and ISO/PAS 21448 ensure that vehicles are able to perform intended functions safely, without failures in electronic systems leading to dangerous situations [1, 2]. They help in managing the safe interaction between software and mechanical components, essential in preventing accidents caused by software errors or hardware failures.

Cybersecurity is another critical area covered by safety standards. As IEVs increasingly rely on software and are connected to networks, they are vulnerable to cyberattacks that could compromise vehicle control systems, steal user data, or even manipulate vehicle functionality. Standards such as ISO/SAE 21434 provide a framework for protecting vehicles against such threats, ensuring data privacy and secure communication for both the vehicle and its connectivity with infrastructure and other vehicles [3].

Environmental safety standards also play a crucial role. These standards ensure that the vehicle's operation contributes minimally to environmental degradation. This includes regulations on emissions and guidelines on the recyclability of vehicle components, particularly batteries, which contain hazardous materials. By complying with these standards, manufacturers help reduce the ecological footprint of their vehicles.

12.1.1. Vehicle Design and Crashworthiness

Safety standards for IEVs include stringent requirements for vehicle design, particularly in terms of crashworthiness. These standards ensure that vehicles can withstand and mitigate the impact of collisions to protect occupants. Standards such as those developed by the National Highway Traffic Safety Administration (NHTSA) in the United States and the European New Car Assessment Programme (Euro NCAP) in Europe include tests for front, side, and rear impacts as well as rollover tests to evaluate a vehicle's safety levels [4, 5, 6, 7].

12.1.2. Battery Safety and Management Systems

Given that most IEVs are powered by lithium-ion batteries, specific standards are in place to manage the risk associated with battery operation, especially under thermal stress, which can lead to thermal runaway and fires. Standards like SAE J2464 provide guidance on electric and hybrid vehicle battery exposure to electrical abuses, thermal management, and reliability [8]. Additionally, the International Electrotechnical Commission (IEC) 62660 series addresses secondary lithium-ion cells for the propulsion of electric road vehicles, including safety requirements related to battery management systems [9].

12.1.3. Software and Cybersecurity

As IEVs often rely on complex software for vehicle control and connectivity, standards such as ISO/SAE 21434 have been developed to address road vehicle cybersecurity, aiming to protect against unauthorized access to vehicle data and systems [3]. This standard covers the entire lifecycle of a vehicle, from design and manufacturing to decommissioning.

12.1.4. Autonomy

For vehicles featuring higher levels of autonomy, standards like ISO 26262 (Road vehicles—Functional safety) and the recently published ISO/PAS 21448 (Road vehicles—Safety of the intended functionality) are crucial [1, 2]. These standards are designed to ensure functional safety and manage risks that could be encountered due to errors in software algorithms, sensor failures, or inadequate performance of the driving system under certain conditions.

12.1.5. Environmental and Electromagnetic Compatibility

IEVs must comply with environmental standards concerning emissions and noise levels. Electromagnetic compatibility (EMC) standards such as SAE J551/5 and ISO 11452 ensure that the vehicle's electronic systems are immune to electromagnetic interference and do not emit excessive electromagnetic waves that could interfere with other devices [10, 11].

12.1.6. Connectivity and Vehicle-to-Everything Communication

As connectivity becomes an integral part of IEVs, standards such as IEEE 802.11p and dedicated short-range communications (DSRC) protocols outlined in IEEE 1609 series are significant [12, 13]. They ensure reliable vehicle-to-everything (V2X) communication, crucial for the safety in autonomous driving scenarios.

Case Study: Minimum IEV Safety Requirements in the United States

In this section, we present a case study on the minimum IEV safety requirements in the United States. IEVs sold in the United States must comply with a series of safety standards established by various regulatory bodies to ensure their safety, reliability, and environmental compatibility. The Federal Motor Vehicle Safety Standards (FMVSS), overseen by the NHTSA, dictate safety requirements across several vehicle components including occupant protection systems like airbags and seat belts, electronic stability controls for maintaining vehicle control, effective braking systems, and lighting and signaling devices to enhance visibility and communication on the road.

Additionally, EVs face unique challenges related to their high-voltage batteries, addressed by specific safety standards such as SAE J2929 and UL 2580. These standards focus on the safety of lithium-ion batteries in electric and hybrid vehicles, emphasizing prevention against fire risks and electric shocks. Despite not emitting exhaust pollutants, EVs are still subject to emissions standards like the Corporate Average Fuel Economy (CAFE) and greenhouse gas emissions standards, which govern the production of their electricity. These standards are crucial for improving fuel economy and reducing emissions across vehicle fleets. Safety standards also extend to the charging systems of EVs. The National Electrical Code (NEC), Article 625, and SAE J1772 standard govern the installation and operation of Electric Vehicle Supply Equipment (EVSE), ensuring safety in the electrical infrastructure used for charging EVs.

As vehicles become increasingly connected, cybersecurity becomes paramount. The ISO/SAE 21434 Road Vehicles—Cybersecurity Engineering standard helps manufacturers protect against cyber threats, safeguarding vehicle data and systems. Compliance with these standards is essential for the sale of EVs in the US market, and failure to adhere can result in recalls, penalties, and loss of certification. These regulations are enforced by federal and state agencies to ensure that EVs provide a safe, reliable, and environmentally responsible mode of transportation.

12.2. Safety of the Intended Functionality

In this section, we delve into the details of the latest autonomy safety standard, Safety of the Intended Functionality (SOTIF), to understand the technical challenges of ensuring safety of autonomous or assisted driving of IEVs [14].

While IEV OEMs race toward more advanced autonomous driving technologies, fatal accidents have continued to happen. In July 2022, a Tesla Model 3 crashed in California and a pedestrian was killed. In August 2022, China's XPeng P7 failed to detect a parked vehicle on a highway while traveling at high speed and crashed directly into the parked vehicle and killed the person standing behind it. The list of fatal accidents caused by IEVs goes on and on, which leads to the belief that we should revisit how we ensure the safety of IEVs powered by advanced autonomous driving technologies.

Functional safety has been the gold safety standard for road vehicles. Functional safety is concerned with mitigating risks due to system failures, including software errors, hardware faults, or sensor failures. Functional safety standards provide guidance on design, verification, and validation measures and define clear methods to categorize a system into different Automotive Safety Integrity Levels (ASILs) [1].

Although the application of functional safety standards helps automotive OEMs achieve safety in situations without failure, these standards do not gauge the system's capability of handling the highly dynamic external environment, such as the system's capability of detecting a pedestrian during heavy rain. Unfortunately, many intelligent EV accidents were caused by the exact reason that the systems were incapable of handling the highly dynamic external environment.

In addition to functional safety, SOTIF focuses on proper situational awareness to be safe. The standard is concerned with guaranteeing the SOTIF in the absence of a fault. For instance, SOTIF should guarantee that the system is capable of detecting obstacles under different situations, or the system knows under which situation it is incapable of functioning properly so that it can warn the driver ahead of time.

However, this is extremely challenging, how can you enumerate all possible situations to test your system? How do you train your AI decision system to handle all possible situations? How do you even predict when your system will fail? For these reasons, enforcing SOTIF in the physical world is nearly impossible, as there is no effective way to encounter all possible situations during physical testing. In contrast, we can conduct an unlimited number of tests under an unlimited number of situations in the virtual world, or digital twin, given that the digital twin is close enough to reality.

The challenge is how to make the digital twin as close to reality as possible, and I have summarized three design principles for the digital twin to be effective in enforcing SOTIF [15, 16] (Figure 12.1).

- Principle 1: structural twin: For realistic simulation, we need a realistic representation of the environment. The digital twin should include precise 3-D models of the environments, which render the same geographical and geometrical properties as the physical counterparts. Based on the structural twin, physical processes of autonomous vehicles can be faithfully modeled. For example, with the geometric information of a target environment, realistic LiDAR point clouds and camera images could be synthesized; with physical road structures and geometric information, navigation and control algorithms can be faithfully verified in the simulator.
- Principle 2: physical twin: Modeling the interactions between physical environments and autonomous vehicles is challenging but imperative for SOTIF. Physical processes of object movements, collisions, and sensing should be faithfully modeled in the digital twin. Physical twin properties should correspond to real-world objects properties to support physical simulation. For instance, a vehicle dynamics model requires a set of parameters to precisely emulate the vehicle's maneuvers; mass and mesh of objects are necessary for modeling collision; texture and color of objects

Figure 12.1 Principles for digital twins to enforce SOTIF.

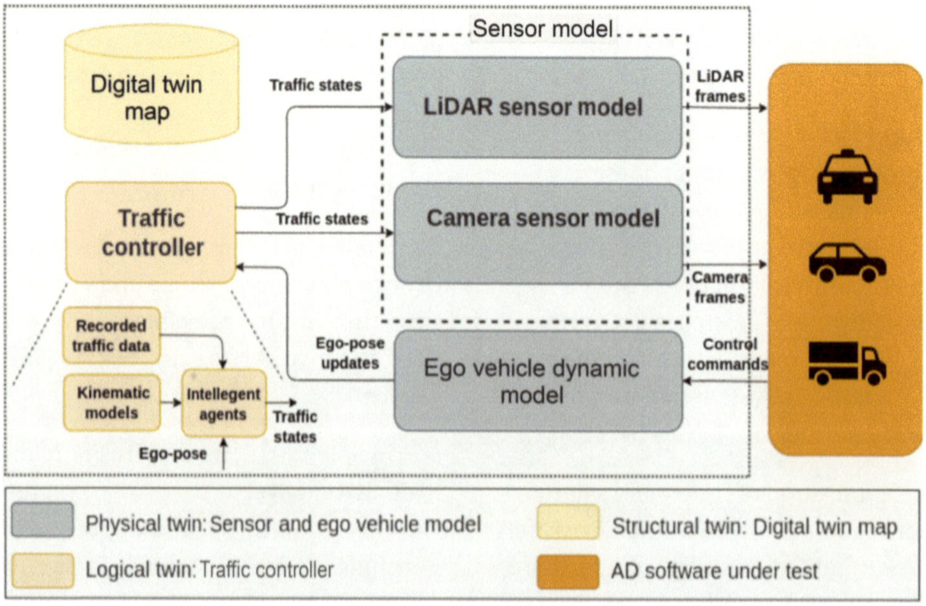

are for rendering camera images and LiDAR point clouds.

- Principle 3: logical twin: The behaviors of other traffic participants, such as pedestrians and other vehicles, need to be modeled as well. Simulated traffic participants (e.g., vehicles and pedestrians) should have a behavior similar to their physical counterparts when interacting with the autonomous vehicle and other objects. This principle is critical for testing the planning and decision module of an autonomous vehicle, especially in scenes of heavy traffic.

With a properly designed digital twin, we can test predefined corner cases to maximize the coverage of autonomous driving algorithms. Furthermore, by modeling the interactions between autonomous vehicles and physical environments as a search problem, we can utilize the digital twin to explore and discover hidden corner cases, further improving the safety coverage.

12.3. Latest Research Progress

In this section, we review two latest research progress on ensuring safety and reliability of autonomous machines such as IEVs.

First, the Vulnerable-Proportional Protection (VPP) design paradigm is an innovative approach designed to enhance the resilience of autonomous machines with lower latency and energy overhead [17, 18]. VPP addresses the need for a tailored protection strategy that diverges from the traditional "one-size-fits-all" methods, which often result in significant performance and cost trade-offs. VPP operates on the principle that protection should be allocated in proportion to the inherent vulnerability of different system components.

Specifically, VPP implements software-based protection strategies for front-end operations, which include sensing, perception, and localization kernels. These front-end components benefit from anomaly detection techniques, leveraging their natural resilience, temporal consistency, and error-masking capabilities. Such an approach significantly reduces unnecessary re-execution overhead and potential protection failures.

For the more critical back-end components, such as planning and control kernels, VPP employs hardware-based protection. This includes modular redundancy and checkpointing to ensure robustness, as these components directly influence vehicle control and require rapid and reliable response capabilities. Evaluative results demonstrate that VPP not only optimizes resilience but does so with minimal increases in latency and energy consumption compared to more conventional protection mechanisms.

This effectiveness is highlighted through testing on autonomous vehicles and drones, showing that VPP can adaptively reduce error propagation rates effectively to zero, fulfilling the highest safety standards like ASIL-D.

Overall, VPP represents a significant step forward in designing protection systems for autonomous machines, promoting both operational efficiency and high safety levels without the compromises typically associated with extensive protective measures.

Second, BRAUM is a dynamic protection system designed to enhance the fault tolerance of autonomous vehicle software with minimal performance overhead [19]. BRAUM explores the inherent error-masking capabilities of various algorithms within the autonomous driving software stack. The core innovation lies in BRAUM's ability to systematically identify these error-masking mechanisms and use them to selectively provide error protection.

BRAUM operates through a large-scale fault injection into the autonomous driving software stack, combined with program analysis techniques to trace error propagation and masking by individual algorithms. The system synthesizes faults that mimic potential issues like transient errors, adversarial attacks, and software bugs, minimizing their impact on the software's behavior. BRAUM's fault injection framework is described as lightweight, preserving the original software stack's functionality while testing.

BRAUM also introduces the notion of a fault tolerance level (FTL), which categorizes algorithms based on how effectively their output errors are masked before reaching the actuators. An iterative algorithm calculates the FTL for each component in the autonomous driving software, which guides the dynamic protection system. This system selectively relaxes protection for algorithms with a high FTL, meaning their errors are likely always masked, thus avoiding unnecessary overhead.

The main contribution of BRAUM is its approach to reducing error propagation rates significantly compared to a baseline without any protection and achieving this with substantially lower execution overhead compared to systems that provide constant, nonselective protection. The results verify BRAUM's effectiveness in significantly lowering the error propagation rates and the system's overhead, making it a practical solution for enhancing the reliability of autonomous driving software with minimal resource expenditure.

12.4. Summary

In summary, with the recent development of the IEV industry, many IEV OEMs have been demonstrating their flashy technologies and positioning themselves as revolutionaries who will completely change the automotive industry, yet the core demand from customers has not changed a bit, safety first. Intelligence, after all, should be elegant instead of ruthless; it is time for the industry to step back a bit from the technology race and reach a consensus on delivering the safest autonomous driving technologies for customers. In this context, SOTIF seems to be the right direction, and digital twin seems to be the right tool.

References

1. International Organization for Standardization, "ISO 26262-1:2018, Road Vehicles—Functional Safety—Part 1: Vocabulary," Standard, 2018.
2. International Organization for Standardization, "ISO/PAS 21448:2019, Road Vehicles—Safety of the Intended Functionality," Publicly Available Specification, 2019.
3. International Organization for Standardization and SAE International, "ISO/SAE 21434:2021, Road Vehicles—Cybersecurity Engineering," Standard, 2021.
4. National Highway Traffic Safety Administration, "FMVSS 208: Occupant Crash Protection," Standard, Revised September 2007.
5. National Highway Traffic Safety Administration, "New Car Assessment Program (NCAP) Testing Protocols," Program Guidelines, Washington, DC, 2021.
6. European New Car Assessment Programme, "Euro NCAP Safety Testing Protocols," Testing Protocols, Brussels, Belgium, 2021.
7. Insurance Institute for Highway Safety, "IIHS Vehicle Safety Ratings Criteria," Methodology Overview, Arlington, VA, 2021.
8. SAE International, "SAE J2464: Electric and Hybrid Electric Vehicle Rechargeable Energy Storage System (RESS) Safety and Abuse Testing," Standard, March 2009.
9. International Electrotechnical Commission, "IEC 62660-1:2018, Secondary Lithium-Ion Cells for the Propulsion of Electric Road Vehicles—Part 1: Performance Testing," 2018.
10. SAE International, "SAE J551/5:2017, Vehicle Electromagnetic Immunity—Off-Vehicle Source," Standard, 2017.
11. International Organization for Standardization. "ISO 11452-2:2013, Road Vehicles—Component Test Methods for Electrical Disturbances from Narrowband Radiated Electromagnetic Energy—Part 2: Absorber-Lined Chamber," Standard, 2013.
12. IEEE, "IEEE 802.11p-2010, IEEE Standard for Information Technology—Telecommunications and Information Exchange between Systems Local and Metropolitan Area Networks—Specific Requirements Part 11: Wireless LAN Medium Access Control (MAC) and Physical Layer (PHY) Specifications Amendment 6: Wireless Access in Vehicular Environments," Standard, July 2010.
13. IEEE, "IEEE 1609.2-2016, IEEE Standard for Wireless Access in Vehicular Environments—Security Services for Applications and Management Messages," Standard, February 2016.
14. Liu, S., "Safety of the Intended Functionality (SOTIF) for Autonomous Driving," Forbes, https://www.forbes.com/sites/forbestechcouncil/2022/09/23/safety-of-the-intended-functionality-sotif-for-autonomous-driving/.
15. Yu, B., Chen, C., Tang, J., Liu, S. et al., "Autonomous Vehicles Digital Twin: A Practical Paradigm for Autonomous Driving System Development," *Computer* 55, no. 9 (2022): 26-34.
16. Yu, B., Tang, J., and Liu, S.S., "Autonomous Driving Digital Twin Empowered Design Automation: An Industry Perspective," in *2023 60th ACM/IEEE Design Automation Conference (DAC)*, San Francisco, 2023, 1-4, IEEE.
17. Wan, Z., Gan, Y., Yu, B., Liu, S. et al., "VPP: The Vulnerability-Proportional Protection Paradigm towards Reliable Autonomous Machines," in *Proceedings of the 5th International Workshop on Domain Specific System Architecture (DOSSA)*, Orlando, 1-6.
18. Wan, Z., Gan, Y., Yu, B., Liu, S. et al., "The Vulnerability-Adaptive Protection Paradigm toward Reliable Autonomous Machines," *Communications of the ACM*.
19. Gan, Y., Whatmough, P., Leng, J., Yu, B. et al., "Braum: Analyzing and Protecting Autonomous Machine Software Stack," in *2022 IEEE 33rd International Symposium on Software Reliability Engineering (ISSRE)*, Charlotte, NC, 2022, 85-96, IEEE.

Chapter 13

Conclusion

This book aims at clarifying complex concepts, making advanced technological topics in the IEV industry understandable for all readers. It not only aids understanding but also guides readers on capitalizing on the monetization opportunities within the IEV market. The inclusion of real-world case studies illustrates the practical application of these theories, enhancing the book's relevance. It also addresses the challenges and obstacles the industry faces, offering a comprehensive overview.

The content starts with an exploration of the monetization opportunities in the IEV sector, highlighting the disruption of traditional automotive business models. Subsequent chapters discuss the global IEV supply chain's evolution, its role in transitioning to an autonomy economy, and its societal impacts. It then covers the innovative approach of scenario-driven car making and the unique project management challenges in IEV development.

Then, this book dives deep into the key technologies of IEVs, including electrification and autonomous driving, enriched with cutting-edge research on battery technologies and AR. It further examines crucial technical areas like semiconductors' role in IEVs, challenges in network communications for cooperative driving, and the stringent safety standards specific to IEVs.

While the book does not cover all aspects of IEVs in depth, those interested in exploring topics not covered in this book can refer to the reading list provided below:

Market Economics and Consumer Adoption

1. D. S. Rapson and E. Muehlegger, "The Economics of Electric Vehicles," NBER Working Paper No. 29093, July 2021, Revised July 2022.—This study discusses the optimal paths for EV adoption, emphasizing the need for policy designs that align private and public economic incentives, suggesting a gradual reduction in purchase incentives as market maturity increases.
2. E. J. Miravete, M. J. Moral, and J. Thurk, "Fuel Taxation, Emissions Policy and Competitive Advantage in the Diffusion of European Diesel Automobiles," *RAND J. Econ.*, vol. 33, pp. 275–297, 2002.—This paper analyzes how fuel taxation and emissions policies have historically influenced vehicle adoption in Europe, offering insights that could inform current EV policy frameworks.
3. J. Kane, "Germany: Plug-In Car Share at 26%: Records Everywhere in December 2020," Insideevs.com, January 16, 2021.—The article reports on Germany's significant increase in plug-in EV adoption, highlighting the effectiveness of supportive policies and infrastructure development.
4. L. M. S. Wappelhorst and P. Mock, "Using Vehicle Taxation Policy to Lower Transport Emissions: An Overview for Passenger Cars in Europe," ICCT, December 2018.—This research provides an overview of how vehicle taxation policies can be strategically utilized to reduce transportation emissions, with a focus on the role of EVs in achieving these goals.
5. "Consumer Adoption of Plug-In Electric Vehicles in Selected Countries," *Future Transportation*, vol. 14, no. 1, pp. 30–45, April 2021.—This study reviews consumer adoption rates of plug-in EVs across various countries, examining the factors that drive or hinder market penetration.

Business Models and Profitability

1. *Electric Vehicle Business Models: Global Perspectives*, Springer.—This book collects insights on new and profitable business models for EVs, focusing on ownership models and total cost of ownership optimizations.
2. A. Coman and A. Wiznizer, "Vertically Integrated Supply Chain of Batteries, Electric Vehicles, and Charging Infrastructure: A Review of Three Milestone Projects from Theory of Constraints Perspective," *Sustainability*, vol. 13, no. 7, p. 3632, March 2021.—The study examines Tesla's vertically integrated supply chain model, contrasting it with earlier EV projects like EV1 and Better Place, to identify production and market bottlenecks and solutions.
3. T. Crispeels, "Profitability Evaluation of Vehicle-to-Grid-Enabled Frequency Containment Reserve Services into the Business Models of the Core Participants of Electric Vehicle Charging Business Ecosystem," *World Electric Vehicle Journal*, vol. 14, no. 1, January 2023.—This research outlines how vehicle-to-grid technology can be integrated into the business models of EV charging infrastructure operators, potentially enhancing profitability.

4. McKinsey & Company, "Improving electric vehicle economics," January 2017.—This report suggests that automakers can enhance the profitability of EVs through innovative business models like battery leasing and targeting fleet sales.

Policy and Economic Incentives

1. Y. Liang, "Policy Incentives for the Adoption of Electric Vehicles across Countries," *Sustainability*, vol. 6, no. 11, pp. 8056–8078, November 2014.—This paper reviews the various policy incentives implemented across different countries to promote EV adoption and discusses the effectiveness of these policies in stimulating the EV market, using the United States as a case study.
2. X. Xu, "Impact of Incentive Policies and Other Socio-Economic Factors on Electric Vehicle Market Share: A Panel Data Analysis from 20 Countries," *Sustainability*, vol. 13, no. 5, p. 2928, March 2021.—This study examines how incentive policies along with socioeconomic factors affect the market share of EVs in 20 different countries, highlighting the positive impacts of tax reductions and charging infrastructure availability.
3. "Global EV Outlook 2023," IEA.—This report discusses the significant policy developments in the European Union and the United States aimed at reducing CO_2 emissions and supporting EV production and adoption, outlining the direct and indirect impacts of these policies on the market.
4. "The Effects of Subsidy Policy on Electric Vehicles and the Supporting Regulatory Policies: Evidence from Micro Data of Chinese Mobile Manufacturers," *Frontiers*.—This article uses microdata to explore the effects of subsidy policies on EV sales in China, utilizing statistical methods to analyze the impact on market dynamics and consumer demand.
5. X. Zheng, "Effect of Policy Incentives on the Uptake of Electric Vehicles in China," *Sustainability*, vol. 11, no. 12, p. 3323, June 2019.—This research focuses on how China's policy incentives, such as subsidies and tax exemptions, have significantly driven the uptake of EVs, particularly analyzing the effects of purchasing and driving restrictions on market growth.

Transportation Systems and City Planning

1. "Electric Vehicle Integration into Road Transportation, Intelligent Transportation, and Electric Power Systems: An Abu Dhabi Case Study," Smart Cities, 2023.—This study evaluates the integration of EVs in Abu Dhabi, analyzing impacts across transportation, intelligent transportation systems, and electric power systems, and suggesting the need for a coordinated transportation-energy system.
2. R. Korzynietz, "Innovation Needs for the Integration of Electric Vehicles into the Energy System," *WEVJ*, vol. 10, no. 4, 2019.—This article discusses the urgent need for intelligent integration of EVs with the energy system, highlighting the challenges and innovations required for smart charging and the use of renewable energies.

3. "Development of Intelligent Electric Vehicles and Smart Transportation," *Energies*, Special Issue, 2024.—This special issue focuses on the technological advancements necessary for the integration of intelligent EVs into smart transportation systems, emphasizing the role of green energy technologies.

4. "Integration of EVs into the Smart Grid: A Systematic Literature Review," *Energy Informatics*, 2023.—This study provides a comprehensive review of research on the integration of EVs into the smart grid, detailing various research methods and focusing on challenges such as power-grid reliability and smart charging.

5. "Sustainable Mobility in Smart Cities: A Document Study of Mobility Initiatives of Mid-sized Nordic Smart Cities," *European Transport Research Review*, 2021.—This study examines the integration of sustainable mobility, including intelligent transport systems and EVs, in the urban development plans of mid-sized Nordic cities, emphasizing a strategic approach to smart city concepts.

6. "New Paradigm of Sustainable Urban Mobility: Electric and Autonomous Vehicles—A Review and Bibliometric Analysis," *Sustainability*, 2021.—The paper reviews the emerging trends of integrating electric and autonomous vehicles in urban mobility, analyzing their potential to enhance sustainability and discussing the challenges of transitioning to these new forms of transportation.

7. "Collaborative Intelligent Traffic Planning in the Internet of Vehicles," *Sensors*, 2024.—This article proposes a collaborative intelligent traffic planning framework using the Internet of Vehicles to address urban congestion through optimized traffic signal timing and vehicle routing, enhancing overall urban mobility efficiency.

8. "Autonomous Road Vehicles: Challenges for Urban Planning in European Cities," *Urban Science*, 2019.— This article discusses the impact of autonomous vehicles on urban planning in European cities, focusing on how these technologies influence land use, infrastructure design, and policy development to support sustainable urban growth.

Supply Chain Management

1. A. Coman and A. Wiznizer, "Vertically Integrated Supply Chain of Batteries, Electric Vehicles, and Charging Infrastructure: A Review of Three Milestone Projects from Theory of Constraints Perspective," *Sustainability*, vol. 13, no. 7, p. 3632, March 2021.—This study analyzes Tesla's vertically integrated supply chain compared to earlier EV projects like EV1 and Better Place, highlighting how Tesla has addressed production and market bottlenecks through a closed-loop supply chain model.

2. "How Tesla Introduced An Intelligent EV Supply Chain," CleanTechnica, March 8, 2023.—This article describes Tesla's evolution in supply chain management, emphasizing how the company has moved from reliance on external suppliers to a more integrated model that includes the development of its computing systems and battery production.

3. R. A. Mancebo Boloy, "Electric Vehicle Supply Chain Management: A Bibliometric and Systematic Review," *Energies*, vol. 16, no. 4, p. 1563, February 2023.—This research presents a comprehensive review of the EV

supply chain, identifying critical materials and potential risks, and discussing trends toward sustainability and transparency in supply chain practices.

4. D. Kalaitzi, A. Matopoulos, and B. Clegg, "Managing Resource Dependencies in Electric Vehicle Supply Chains: A Multi-tier Case Study," *Supply Chain Management*, vol. 24, no. 2, pp. 256–270, January 2019.—This paper explores how EV manufacturers manage resource dependencies with suppliers during the ramp-up phase of production, using a multitier case study approach to illustrate the shift toward more relational intercompany relationships.
5. "Electric vehicles and supply chain: PwC," PwC, Accessed 2024.—PwC's analysis highlights the challenges and opportunities in the EV supply chain, noting the shift from ICEs to EVs and the resulting impacts on suppliers of traditional automotive components.

Product Design

1. "Intelligent Electric Vehicle Control, Testing and Evaluation," *WEVJ*, Special Issue.—This special issue explores various aspects of IEV design, including dynamic simulation, control of automotive intelligent chassis, and innovations in vehicle dynamics control and electric drive assembly.
2. "A Review on Electric Vehicles: Technologies and Challenges," *Smart Cities*, vol. 4, no. 1, 2021.—This review article discusses advancements in EVs, focusing on battery technology, charging methods, and new research challenges, particularly highlighting the integration of advanced technologies in vehicle design.
3. "Trends in electric-vehicle design," McKinsey & Company, 2021.—This report from McKinsey & Company discusses current trends in EV design, emphasizing how OEMs are making strategic choices regarding battery and thermal-management design to optimize cost and performance.
4. "Research Progress on Intelligent Electric Vehicles," *Sensors*, Special Issue.—This special issue addresses the progress in the field of IEVs, focusing on advanced sensors and safety control for connected and automated vehicles.
5. "The future of car interiors in automotive," McKinsey & Company, 2021.—This article provides insights into the evolution of vehicle interiors, discussing how changes in EV design will likely lead to innovative features in connectivity and HMI to improve the in-car experience.
6. "Innovative Motor Designs for Electric Cars," *IEEE Spectrum*, 2021.—This article highlights innovative approaches in the design of electric motors for EVs, emphasizing the importance of optimization techniques and collaboration in developing high-performance motors.
7. "Actuators for Intelligent Electric Vehicles," *Actuators*, Special Issue.—This special issue focuses on advanced actuators in IEVs, exploring their role in enhancing vehicle safety, comfort, and maneuverability, with topics including antilock braking systems and autonomous driving technologies.

Environmental Impacts

1. "Electric Cars: Technical Characteristics and Environmental Impacts," *Environmental Sciences Europe*, 2021.—This article reviews the energy efficiency and environmental impacts of battery EVs, particularly focusing on their advantages over ICE vehicles in reducing emissions and energy consumption.

2. "Primary Energy Use and Environmental Effects of Electric Vehicles," *World Electric Vehicle Journal*, 2021.—This article analyzes the primary energy needs for EV propulsion and discusses the environmental benefits, particularly in terms of CO_2 emission reductions when using renewable and nuclear energy sources in national electricity grids.
3. *Electric Vehicles for Environmental Sustainability*, Springer, 2021.—The book discusses the broader sustainability challenges and opportunities presented by EVs, including impacts on urban air quality and reductions in greenhouse gas emissions.
4. "Estimating the Environmental Impacts of Global Lithium-Ion Battery Supply Chains," *Sustainability*, 2021.—This article provides a detailed analysis of the cradle-to-gate environmental impacts of nickel–manganese–cobalt and lithium–iron–phosphate batteries, emphasizing the need for sustainable practices in the global supply chain.
5. "Environmental Impacts of Hybrid and Electric Vehicles—A Review," SpringerLink, 2021.—This article reviews existing studies on the environmental impacts of hybrid and EVs, focusing on lifecycle assessments to compare global warming potential and other environmental impacts against ICEs.

Project Management

1. "The new BMW: business model innovation transforms an automotive leader," Emerald Insight.—This article discusses BMW's shift toward integrating technology and customer services into their core operations, illustrating how business model innovation is being applied within the automotive industry to adapt to changing environments.
2. "Building smarter cars with smarter factories: How AI will change the auto business," McKinsey & Company.—This resource explores the influence of AI on the automotive industry, particularly how AI will revolutionize project management, from improving production processes to enhancing supply chain efficiency and R&D prioritization.
3. "Tesla and the Reshaping of the Auto Industry," *Management and Organization Review*, Cambridge Core.—The article provides an in-depth look at how Tesla's approach to car manufacturing, including its focus on EVs and integration of high technology, is reshaping traditional automotive industry standards and project management practices.
4. "Big data in real life: The impact of analytics on car manufacturing," SAS.—This article details how big data analytics is being utilized in the car manufacturing process to optimize everything from supply chains to customer relationship management, significantly impacting project management within the industry.
5. "Tesla's Organizational Structure (An Analysis)—Panmore Institute."—This analysis explores Tesla's unique organizational structure, highlighting its flat hierarchy and how it facilitates rapid innovation and agility in the car manufacturing process.
6. "Ford Motor Company's Organizational Structure Characteristics," Panmore Institute—This article details the organizational structure of Ford Motor Company, focusing on its global operations and functional groups, which support its extensive manufacturing and marketing activities.

Index

About the Author

Dr. Shaoshan Liu

Currently Dr. Shaoshan Liu is the Director of Embodied Artificial Intelligence at Shenzhen Institute of Artificial Intelligence and Robotics for Society (AIRS). For the past decade, he has been a pioneer and technical leader in the intelligent electric vehicle industry. He has served as the chief scientist of BeyonCa, a premium electric vehicle company, and is the founder of PerceptIn, an autonomous driving technology company. He was also a senior architect and founding member of Baidu's autonomous driving team.

In addition, Dr. Liu is the founder of the Institute of Electrical and Electronics Engineers (IEEE) Special Technical Community on Autonomous Driving Technologies, Chair of IEEE Entrepreneurship Asia, Chair of IEEE International Roadmap of Devices and Systems (IRDS), a member of Association of Computing Machinery (ACM)'s US Technology Policy Committee, a member of US National Academy of Public Administration's (NAPA) Technology Leadership Panel Advisory Group, a member of the Forbes Technology Council, and an agenda contributor of the World Economic Forum (WEF).

Dr. Shaoshan Liu is also a member of the Global Young Academy, an IEEE Senior Member, an IEEE Distinguished Speaker, an ACM Distinguished Speaker, and a member of American Association for the Advancement of Science (AAAS). His educational background includes a B.S. in Computer Engineering, a M.S. in Biomedical Engineering, a Ph.D. in Computer Engineering from the University of California, Irvine, and a Master of Public Administration (M.P.A.) from Harvard University.

In the field of intelligent electric vehicles, Dr. Liu has published more than 100 research papers and holds more than 150 patents. He has also published three textbooks on autonomous driving technologies:

1. Shaoshan Liu et al., *Robotic Computing on FPGAs*, Synthesis Lectures on Computer Architecture, Morgan & Claypool, 2021.
2. Shaoshan Liu, *Engineering Autonomous Vehicles and Robots: The DragonFly Modular-Based Approach*, IEEE Wiley, 2020.
3. Shaoshan Liu et al., *Creating Autonomous Vehicle Systems*, Synthesis Lectures on Computer Science, Morgan & Claypool, 2020.

www.ingramcontent.com/pod-product-compliance
Lightning Source LLC
LaVergne TN
LVHW070953210926
848482LV00002B/197
9781468608496